Hazards of Nitrogen and Other Inert Gases

Hazards of Nitrogen and Other Inert Gases

How They Can be Safely Managed

M. Darryl Yoes
Safety Consulting International, LLC
Jackson, LA, USA

Copyright © 2025 by John Wiley & Sons Inc. All rights reserved, including rights for text and data mining and training of artificial intelligence technologies or similar technologies.

Published by John Wiley & Sons, Inc., Hoboken, New Jersey.

Published simultaneously in Canada.

No part of this publication may be reproduced, stored in a retrieval system, or transmitted in any form or by any means, electronic, mechanical, photocopying, recording, scanning, or otherwise, except as permitted under Section 107 or 108 of the 1976 United States Copyright Act, without either the prior written permission of the Publisher, or authorization through payment of the appropriate per-copy fee to the Copyright Clearance Center, Inc., 222 Rosewood Drive, Danvers, MA 01923, (978) 750-8400, fax (978) 750-4470, or on the web at www.copyright.com. Requests to the Publisher for permission should be addressed to the Permissions Department, John Wiley & Sons, Inc., 111 River Street, Hoboken, NJ 07030, (201) 748-6011, fax (201) 748-6008, or online at http://www.wiley.com/go/permission.

Trademarks: Wiley and the Wiley logo are trademarks or registered trademarks of John Wiley & Sons, Inc. and/or its affiliates in the United States and other countries and may not be used without written permission. All other trademarks are the property of their respective owners. John Wiley & Sons, Inc. is not associated with any product or vendor mentioned in this book.

Limit of Liability/Disclaimer of Warranty: While the publisher and author have used their best efforts in preparing this book, they make no representations or warranties with respect to the accuracy or completeness of the contents of this book and specifically disclaim any implied warranties of merchantability or fitness for a particular purpose. No warranty may be created or extended by sales representatives or written sales materials. The advice and strategies contained herein may not be suitable for your situation. You should consult with a professional where appropriate. Further, readers should be aware that websites listed in this work may have changed or disappeared between when this work was written and when it is read. Neither the publisher nor authors shall be liable for any loss of profit or any other commercial damages, including but not limited to special, incidental, consequential, or other damages.

For general information on our other products and services or for technical support, please contact our Customer Care Department within the United States at (800) 762-2974, outside the United States at (317) 572-3993 or fax (317) 572-4002.

Wiley also publishes its books in a variety of electronic formats. Some content that appears in print may not be available in electronic formats. For more information about Wiley products, visit our web site at www.wiley.com.

Library of Congress Control Number: 2025905850
Hardback ISBN: 9781394330263
ePDF ISBN: 9781394330287
epub ISBN: 9781394330270

Cover Design: Wiley
Cover Images: © zorazhuang/Getty Images, US Chemical Safety and Hazard Investigation Board (CSB)

Contents

Preface *ix*
Note to the Reader *xiii*
About the Author *xv*

1 The Properties, Uses, and Safety Hazards of Nitrogen *1*

2 The Properties, Uses, and Safety Hazards of Other Inert Gases *13*
2.1 Argon (Ar) *13*
2.2 Carbon Dioxide (CO_2) *15*
2.3 Carbon Monoxide (CO) *16*
2.4 Helium (He) (Overview) *17*
2.5 Neon (Ne) *19*
2.6 Krypton (Kr) *20*
2.7 Xenon (Xe) *21*
2.8 Light Hydrocarbons *21*

3 The Effects of Nitrogen and Other Asphyxiants on the Body (Oxygen Deprivation) *25*

4 Protection for Personnel Against Inert Gas Asphyxiation and/or Cold Burns *31*
4.1 Adequate Warning Signs and Barricades *31*
4.2 Personnel Gas Monitors and Continuous Gas Quality Monitoring and Alarms *33*
4.3 Use of Self-Contained Breathing Apparatus *35*
4.4 The Documented Work Permit (The Authorization for Specified Work to Begin Issued by an Authorized and Designated Individual) *36*
4.5 The Union Carbide Nitrogen Asphyxiation Incident – Date of Incident 27 March 1998 Investigated by the Chemical Safety and Hazard Investigation Board Final Report Issued 23 February 1999 (Report Number 98-05-I-LA) *36*
4.6 Confined Space Entry *38*
4.7 Protection Against Supercold Liquids such as Liquid Nitrogen *39*

5	**Confined Space Entry – The Occupational Safety and Health Administration Standard (29 CFR 1910.146) and Some Key OSHA 'Letters of Interpretations'** *45*
5.1	The US OSHA Confined Space Regulation '29 CFR 1910.146 – Permit Required Confined Spaces' *45*
5.2	Confined Space Entry Letters of Interpretation by OSHA (Standard 29 CFR 1910.146) *64*

6	**The Hazard of Contaminated Breathing Air and How It Can Kill** *95*
6.1	The Use of Utility Air or Instrument Air as Breathing Air Should Not Be Permitted *95*
6.2	Fatalities Due to Using Blended or Manufactured Breathing Air *96*
6.3	Requirements for Breathing Air to Ensure Quality (from the OHSA Technical Manual on Respiratory Protection) *96*
6.4	Summary of OSHA Requirements for Breathing Air Quality *97*
6.5	Notes from the Regulation and the OSHA Technical Manual on Respiratory Protection *97*
6.6	Other Specific Requirements *98*

7	**Most Frequent Causes of Nitrogen Asphyxiation and How to Address Them** *103*

8	**More on Safe Utility Connections** *111*

9	**The Hazards of Inert Entry and an Overview of the Process (Includes Case Studies of What Has Happened)** *117*
9.1	Background *118*
9.2	Specialized Inert Entry Procedures *119*
9.3	Inert Entry Planning *120*
9.4	The Job Safety Analysis *121*
9.5	Acceptable Inert Atmosphere *121*
9.6	Inert Gas Supply and Quality *122*
9.7	Breathing Air System *122*
9.8	Assuring Breathing Air Quality *125*
9.9	The Contractor Selection Process *126*
9.10	Life Support Equipment *127*
9.11	Confined Space and Inert Entry Rescue Plan *127*
9.12	Video Surveillance Equipment/Rescue Equipment *127*
9.13	Reactor Preparation for Entry *128*
9.14	Caution *128*
9.15	Catalyst Crusting (Another Caution) *129*

10	**Carbon Capture, Use, and Storage** *133*
10.1	How is the Industry Responding to This Increase in Carbon Dioxide? *134*
10.2	Safety Aspects of Carbon Capture, Use, and Storage (CCUS) *137*

11	**Nitrogen Asphyxiation Case Studies, and Asphyxiation by Other Inert Gases** *143*	
11.1	What Has Happened / What Can Happen Incident Case Study Number 1 – Nitrogen Asphyxiation *144*	
11.2	What Has Happened / What Can Happen Incident Case Study Number 2 – Nitrogen Asphyxiation *148*	
11.3	What Has Happened / What Can Happen Incident Case Study Number 3 – Argon Asphyxiation *152*	
11.4	What Has Happened / What Can Happen Incident Case Study Number 4 – Argon Asphyxiation *155*	
11.5	What Has Happened / What Can Happen Incident Case Study Number 5 – Carbon Dioxide Asphyxiation *159*	
11.6	What Has Happened / What Can Happen Incident Case Study Number 6 – Low Oxygen Content (Asphyxiation) *160*	
11.7	What Has Happened / What Can Happen Incident Case Study Number 7 – Employee Dies due to Asphyxiation from Oxygen Displacement *161*	
11.8	What Has Happened / What Can Happen Incident Case Study Number 8 – An Explosion Occurred While Unloading Liquid Nitrogen at an Ice Cream Facility (10 Injured) *163*	
11.9	What Has Happened / What Can Happen Incident Case Study Number 9 – Fatality of Welder in Confined Space Welding in the Presence of Argon *163*	
11.10	What Has Happened / What Can Happen Incident Case Study Number 10 – A Summary of Incidents Involving 'Would-be Rescuers' *164*	
12	**Summary of Additional Actions to Help Prevent Asphyxiation Incidents at Our Facilities** *171*	
13	**Additional Discussion of Liquid Nitrogen Use in Ice Cream Shops** *175*	

End of Book Quiz *181*
Appendices: *185*
    1 Answers to the End of Chapter Quizzes *185*
    2 Answers to the End of Book Quiz *215*
    3 Documented Incidents Involving Nitrogen Resulting in Fatalities or Serious Injury *219*
Index *251*

Preface

My wife and I recently took a short vacation to South Florida, visiting the Everglades and Key West. It was a great chance to get away for a few days and enjoy the beautiful scenery and sun. While passing through Key Largo, we stopped briefly at an ice cream shop, and to my surprise, I discovered 'Nitrogenated Ice Cream.' For those familiar with it, it will be no surprise that the shop had a 'selfie station' where customers could snap a picture of themselves directly in front of their liquid nitrogen storage tank. An image of the liquid nitrogen tank located inside the ice cream parlour is on the next page. You will note that it is labelled as a 'selfie station' where customers are invited to take photographs of themselves and others.

I was painfully aware of the recent investigation report released by the US Chemical Safety Board of six fatalities that had occurred at the Foundation Food Group (FFG) facility in Gainesville, Georgia, due to asphyxiation by liquid nitrogen. This incident occurred on 28 January 2021, when the liquid nitrogen level rose and overflowed in an immersion freezer, releasing the liquid into a confined space, the room where the freezer was located.

This book describes this incident in more detail as a case study. The six deaths were due to oxygen deprivation, the most significant hazard associated with releasing nitrogen into confined areas. Two of these deaths occurred due to the release of liquid nitrogen. The other four deaths occurred when other employees attempted to investigate the release or rescue of the first two employees. In my experience, this is very common in nitrogen incidents due to the human urge to help others in trouble. Rarely is only one person killed by nitrogen. Frequently, another person, and sometimes several others, also die while attempting to render aid to those who are already down. This is precisely what happened at the Foundation Food Group facility.

A similar incident occurred in April 2013 at a beer brewery in Mexico where seven workers, including three company employees, were killed in a confined space entry accident while cleaning a beer fermentation tank. It was believed this incident occurred due to inert gas, either carbon dioxide or nitrogen, and all three of the employees died while attempting to rescue the four downed contractor employees.

So, while at the ice cream shop, I attempted to understand more about using liquid nitrogen in 'nitrogenated ice cream.' One of the high school-age workers explained how they use liquid nitrogen to freeze the mixture of dairy ingredients to make the ice cream. Please note that liquid nitrogen is stored in a cylinder located inside the shop under pressure (250–350 PSI). When released, the very cold liquid, at temperatures of –320 °F (–196 °C), is released into the environment. The liquid nitrogen cylinder was inside the shop, directly adjacent to the serving counter. I asked about safety, and she explained that 'if the alarm goes off, they are supposed to go outside.' Note how she explained that 'they are supposed to go outside'; she did not say they immediately go outside.

This is the issue: when liquid nitrogen is released into a confined space, for example, into the ice cream shop, the nitrogen quickly displaces the oxygen in the room and can result in people being killed due to oxygen deprivation. Due to nitrogen's completely colourless and odourless properties, the victims are unaware of what is happening to them. In this book, you will see just how quickly this can happen and, unfortunately, how frequently it happens and how deadly it is.

This is exactly what happened at the Foundation Food Group, and six people were killed by hypoxia due to oxygen deprivation. Several others were hospitalized due to the effects of nitrogen exposure. This incident is covered in more detail in the What Has Happened / What Can Happen section of this book.

I asked the shop attendant if they had ever had a nitrogen release, and she explained, 'Yes, just two weeks ago, a pipe broke on the tank, and the nitrogen was released.' She said that the nitrogen (supercold) removed the lettering previously on the front lower part of the dairy counter, and sure enough, you could see that the lettering was smudged and essentially gone. She explained that this incident happened overnight when no one was in the shop.

After I returned home, I attempted to contact the business owner of the ice cream shop several times to no avail. Each time I call, I get a recording on the phone, something to the effect that this device is not set up to receive messages. I haven't given up and would certainly like to discuss this with the business owner and hear a little more about how safety is managed at this location. I was in the Nashville area leading a safe operations training course recently and mentioned my experience at this ice cream shop. I was surprised to hear from several attendees that there are shops of this type in the Nashville area, and I have since learned that there are also shops of this type in New Orleans and other major cities across the United States. This 'nitrogenated ice cream' appears to be a national trend. I have since learned of other nitrogenated alcoholic drinks, taking advantage of the refrigeration qualities of liquid nitrogen and the aesthetic value of 'smoking drinks.'

The stainless-steel tank is the compressed liquid nitrogen storage tank. The lettering on the tank says, 'Selfie Station,' where kids and others can have their pictures taken in front of it.

The caption says:

'SELFIE STATION – SPARKY LOVES SELFIES..... TAKE YOURS HERE AND TAG US TO BE FEATURED ON OUR ACCOUNT. SPARKY WILL RANDOMLY PICK POSTS THAT HE LOVES AND SEND YOU A GIFT!!'

Ice cream shop interior: Key Largo, Florida.
Note that the lettering is missing from directly below the counter.

The lettering was removed by a release of supercold liquid nitrogen a couple of weeks before when a section of pipe associated with the nitrogen storage tank failed overnight. When I saw this and coupled it with my knowledge of the Georgia poultry processing facility where six workers were killed by nitrogen, this became the driver for trying to share these lessons learned and hopefully prevent another tragic incident from occurring.

This became the incentive to write this book and help get the message out about the safety aspects of working with or around nitrogen and other asphyxiants in any form. All nitrogen (and other asphyxiants) has the potential for the same hazards, including compressed nitrogen, liquefied nitrogen and pipeline-supplied nitrogen. They are all deadly because if released into a confined space, the nitrogen displaces oxygen, which is a hazard. One breath of nearly pure nitrogen and your brain shuts down, and even if you fall into clean air, you may not start breathing unless you are resuscitated. This is the severe hazard of nitrogen. Liquid nitrogen is also more likely to freeze your skin, resulting in frostbite or possibly severe cold burns due to the extremely cold temperature. Ingestion of even very small amounts of liquid nitrogen can destroy a victim's internal organs. Other gases such as carbon dioxide (frequently used as a fire suppressant), argon (frequently used for welding) and helium, among others, have very similar characteristics and hazards.

In this book, I cover several cases of nitrogen asphyxiation and asphyxiation by other inert gases. If you work with or near equipment containing nitrogen or other inert gases, I encourage you to become familiar with these hazards. Nitrogen and other inert gases can be managed safely, but it only takes one slip or omission for people to be killed.

I later learned that a liquid nitrogen release occurred at a similar ice cream shop in Weston, Florida, where one worker collapsed unconscious, and two responders, a firefighter and a deputy sheriff, were also overcome. The news video of this incident shows the glass shop windows completely iced over. One person was admitted to the hospital for treatment, and the shop was closed for several days.

In addition to the personnel safety hazards created by using nitrogen as a refrigerant in enclosed spaces, ice cream made with liquid nitrogen has resulted in severe injuries due to ingesting small amounts of liquid nitrogen. I'll talk more about this in the last chapter. Still, serious injuries to the digestive tract have occurred due to supercold nitrogen and the rapid expansion of the liquid nitrogen to vapour inside the digestive tract. As a result, at least one US agency has advised against consuming ice cream made with liquid nitrogen.

While I enjoy occasional ice cream, I can't support this trend. I don't support having a liquid nitrogen storage tank indoors where people, including children, are present. Knowing what has happened to others, I am also wary of the possibility of ingesting even a small amount of liquid nitrogen from eating something as delicious as ice cream.

In this book, we will spend most of our time reviewing the background of using nitrogen in industry and elsewhere, the hazards of working with and around nitrogen, and case studies associated with incidents involving nitrogen and nitrogen asphyxiation. Due to the hazards of other asphyxiants and the similarities with nitrogen, we will also discuss several other inert gases and cover several case studies where they were involved. Near the end of the book, there is also a summary of US Occupational Safety and Health Administration (OSHA) NITROGEN ASPHYXIATION INCIDENTS from the OSHA database that I think you will find interesting and useful. You may notice a little redundancy or a repeat of some information as you read this material. This is by design, and you will understand why when you realize the number of deaths that have occurred and how quickly it can happen.

You will also find an End of Chapter Quiz at the end of each chapter as a reminder of the important 'learnings' in the chapter and an End of Book Quiz near the end of the book. The answers to each quiz are available in the appendix. I encourage you to do a quick review using this guide to help you ensure that you have picked up on the important points made.

I hope you enjoy the book and have an opportunity to learn something about working with and around nitrogen and other inert gases and the safety aspects that go along with it. I strongly encourage you to share lessons learned with your peers. You may just save a life by doing so.

The photograph on the cover of this book was provided courtesy of the US Chemical Safety and Hazard Investigation Board (CSB).

Note to the Reader

The author has over 50 decades of experience in the petroleum industry, with operational experience in nearly all aspects of refining operations. Formally an operator and manager, including the role of Safety and Environmental Manager at a major US Gulf Coast refinery, he has led the process of safe operations training worldwide for petroleum refineries and petrochemical plants for the past nearly 20 years.

It is believed that the information provided in the book will help lead to improved safety performance in the petroleum and petrochemical industry and all other industries that use nitrogen or other inert gases. However, neither Safety Consulting International L.L.C., the author, M. Darryl Yoes, nor the company producing the documents contained herein warrants or represents, expressly or by implication, the correctness or accuracy of the content or the information presented here. This material is presented to improve operations safety worldwide and is not intended to be used as operating guidance or procedures. Any use of the material contained herein is done with the user accepting legal liability or responsibility for the consequence of its use or misuse.

Safety Consulting International, LLC, the author, or any other company or person, as outlined earlier, makes no claim, representation or warranty, expressed or implied, that acting in accordance with this book or its contents will produce any particular results with regard to the subject matter contained herein, or satisfy the requirements of any applicable federal, state or local laws and regulations. Nothing in this document constitutes technical advice. If such advice is required, it should be sought from an attorney, a qualified company or a qualified individual.

No material known to be copyrighted or proprietary has been intentionally used in this document without the owner's express approval or permission. The author has made extraordinary efforts to contact all owners of images used in this book to secure consent to use the photographs. All information presented is readily available in the public sector or represents the author's experience and over 50 years of collecting process safety information. The copyright owner approved most images; those without attribution were taken by the author. There are several where, after an extensive search, the original could not be found; these are noted as such.

About the Author

M. Darryl Yoes is a refinery supervisor and manager with practical experience, having spent over 50 years in the refining industry. He began his career as a Process Apprentice at one of the largest US Gulf Coast refineries. After quickly advancing through several supervisory and management assignments, including first and second-line supervisor positions in process, mechanical and technical, he advanced to the position of Risk Management Advisor. He then became the Safety, Health, and Environmental Manager of one of the largest petroleum refineries in the world.

He either led or participated in numerous process safety incident investigations, including those involving the tragic loss of life. He served as an emergency coordinator and incident manager in the role of refinery superintendent and has extensive experience managing other emergency operations, including oil spill response operations. He has also led or participated in numerous process hazard analysis studies, as well as in operations integrity and safety risk assessments at major US and European petrochemical sites. He has also served as the Safety Manager for major construction projects at US refineries.

Mr. Yoes continues process safety management consulting for refining and petrochemical plants and for construction safety management for major construction projects across the United States and is a course leader and facilitator for petrochemical plant's Operations Safety Management training designed for refinery and petrochemical plant management, supervision, engineering and plant operators. He has led comprehensive process safety operations training courses in essentially all regions of the world, including North and South America, Asia, and Europe, to share lessons learned from industry process safety incidents involving loss of containment of flammable or toxic materials leading to fire, explosions and injuries or fatalities. The Safe Operations Training presents practical methods or work practices to help avoid a reoccurrence of these types of incidents.

He is the author of another Wiley textbook and online process safety resource titled 'Process Operations Safety – The What, Why, and How Behind Safe Petrochemical Plant Operations,' which is available at most bookstores, Amazon, and online. This book is a comprehensive overview of process safety in petroleum refineries and petrochemical plants and is a valuable resource to operators, plant supervisors, engineers, and others responsible for process operations safety. To a significant extent, the content of the Safe Operations Training course is reflected in this book.

Mr. Yoes is also the founder and President of Safety Consulting International, LLC, which developed this helpful guide to prevent process safety incidents. He is employed by EcoScience Resource Group, LLC, in Baton Rouge, a firm specializing in developing process operations training manuals, process operating and environmental procedures, and operator training guides for the petroleum refining and petrochemical industries.

1

The Properties, Uses, and Safety Hazards of Nitrogen

The six fatalities that occurred at the Foundation Food Group facility during January 2021, coupled with my visit to the ice cream shop in Florida in early 2024, with the liquid nitrogen tank located inside the shop, made me realize how little the population understands the hazards of nitrogen and other inert gases. Nitrogen is perfectly safe when stored and used so that release or loss of containment is prevented in an enclosed space or other areas where personnel are present and ventilation is limited. These gases are inert. However, exposure to concentrated mixtures with air can be and has been deadly. This book will provide more details on what has happened and what can happen when inert gases like nitrogen are released. We will discuss the hazards of these gases, how they can affect the human body and how this contact can be avoided. We will also review several case studies where this has occurred and, unfortunately, how people have died when this occurs.

The US Environmental Protection Agency (EPA) Cameo Database tells us that nitrogen is a colourless and odourless gas. It is non-combustible and non-toxic. Nitrogen makes up the major portion of the atmosphere (78%), but it does not support life by itself. We also know that nitrogen is frequently used in food processing, purging air conditioning and refrigeration systems. Nitrogen can result in asphyxiation by displacement of oxygen in the air. Also, pressurized nitrogen cylinders under prolonged exposure to fire or heat may rupture violently and become projectiles.

Nitrogen is readily available in the environment and makes up 78% (by volume) of the air we breathe. The remainder is 21% oxygen and a minimal amount of argon. Nitrogen is a non-poisonous gas; however, people and animals can die due to nitrogen concentration. They die not from nitrogen but due to oxygen deprivation. Nitrogen displaces oxygen from the air we breathe. In other words, when the nitrogen concentration increases, the oxygen concentration decreases.

Nitrogen is readily available by cryogenic fractionation of air to extract it for industry and other uses. It is slightly lighter than air and is slightly soluble in water. Liquid nitrogen boils at −320 °F (−196 °C) and contact with it can cause significant cold burns if it touches the skin or flesh.

One volume of liquid nitrogen expands to approximately 700 volumes of gas. An unplanned release of liquid nitrogen, for example, from a liquid nitrogen tank, will fill a standard-size room with concentrated nitrogen in seconds. People in the room can be quickly overcome by oxygen deprivation, and they can die, and this can happen in minutes. Cold nitrogen is heavier than air, and it will accumulate at ground level during a release. When liquid N_2 is exposed to air, the cloudy vapour you see is only the condensed moisture from the air, not the N_2 gas. Remember, nitrogen gas is invisible and odourless, and this is the danger!

Nitrogen condenses at its boiling point, −320.4 °F (−195.8 °C), to a colourless liquid lighter than water (liquid nitrogen). Liquid nitrogen is stored in specially designed pressure vessels designed to protect the super low temperature of the stored cryogenic liquid. When withdrawn, the temperature is −320 °F (−195 °C), making it very useful as a coolant or refrigerant. This characteristic makes liquid nitrogen useful as a refrigerant for food processing, pharmaceutical manufacturing and other commercial uses.

Nitrogen is non-flammable and does not support combustion. Therefore, it is often considered an inert gas (although it is not truly inert).

Safety hazards of nitrogen:
There are three primary hazards associated with nitrogen. In either form, as a liquid or gas, the deadly characteristic is that nitrogen quickly displaces oxygen from the environment. This has resulted in many incidents of accidental asphyxiation of workers. As stated earlier, it is not the nitrogen that kills; people die due to oxygen deprivation when nitrogen displaces oxygen. See the case studies in this book and the appendix for a discussion of nitrogen-related incidents.

Nitrogen (A frequent cause of fatalities and the primary subject of this book – this is a summary; refer to the Safety Data Sheet (SDS), or sometimes referred to as the Material Safety Data Sheet (MSDS) for additional information):
A general description of nitrogen from the US Environmental Protection Agency (EPA) Cameo Chemicals Database:

Nitrogen is a colourless and odourless gas. It is non-combustible and non-toxic and makes up the major portion of the atmosphere, but it does not support life by itself. Nitrogen may cause asphyxiation by displacement of air containing oxygen, which is needed to sustain life. Under prolonged exposure to fire or heat, containers may rupture violently and rocket.

Health hazards of nitrogen from the EPA Cameo Chemicals Database and the Safety Data Sheet:
Vapours from liquefied gas may cause dizziness or asphyxiation without warning. The victim may not be aware that they are being overcome. Vapours from liquefied gas are initially heavier than air and spread along the ground.

High concentrations of nitrogen may cause asphyxiation. Symptoms may include loss of mobility and loss of consciousness. The victim may not be aware of asphyxiation. To protect yourself, wear a self-contained breathing apparatus (SCBA) when responding to someone who has been overcome. Remove the victim to an uncontaminated area and keep the victim warm and rested. Call a doctor. If breathing ceases, apply artificial respiration. Additional hazards associated with exposure to oxygen deprivation are highlighted in Chapter 3 including a chart highlighting the effects of low oxygen concentrations on the body.

If you are working near a vessel that is being purged with nitrogen, appropriate warnings and barricades must be in place to prevent personnel from entering areas that may have an oxygen deficiency. Barricades must be designed to keep personnel from entering these areas, including all areas where nitrogen is vented into the atmosphere.

Personnel should be prevented from entering platforms near open access plates or on process vessels with nitrogen venting. Personnel working near nitrogen venting must be properly outfitted with either hose line-supplied breathing air or protected by a self-contained breathing apparatus. Additionally, personnel while wearing self-contained breathing air should continuously monitor other workers who are working in or near an area where nitrogen purge is in progress. No one should be allowed on the platform or near the vessel opening where nitrogen is venting without breathing air.

Warning: As a reminder, a cartridge or chemical respirator is not sufficient for breathing protection when exposed to an oxygen-deficient atmosphere. The respirator must have a supply of hose-line-supplied or self-contained breathing air.

Do I have to be inside a confined space to be overcome by nitrogen?

> NO, you do not! There have been several cases where workers have been outside a vessel being purged with nitrogen and collapsed due to oxygen deprivation and fell into the vessel and died, or they were overcome, fell and died from the fall.

The second hazard is associated with liquid nitrogen. Liquid nitrogen is a cryogenic liquid and is stored in a pressure vessel specially designed to maintain liquids at extremely low temperatures. It is stored at a temperature of −320 °F (−196 °C) and is an extreme risk of causing cold burns. A cold burn is no different from a hot liquid or a thermal burn. It can result in damage to the skin and to the underlying tissue and can require skin grafts and treatment, not unlike a thermal burn.

When liquid nitrogen is released into the atmosphere, it quickly forms a white fog by freezing the moisture in the air. It may also freeze anything nearby and can create an oxygen-deficient atmosphere.

The third significant hazard with liquid nitrogen is the potential ingestion of even a small amount of the liquid, for example, in ice cream or a cold treat. I'll discuss this in more detail in future chapters.

Humans must breathe oxygen to survive and suffer adverse health effects when the oxygen concentration drops below 19.5%. This can happen when the oxygen in the work environment is displaced by an inert gas such as nitrogen, argon, carbon dioxide or another inert gas. Some light hydrocarbons, although not inert, may have similar effects.

The US Occupational Safety and Health Administration (OSHA) has developed a nice chart that clearly details the effects of an oxygen-deficient atmosphere on personnel who may be unaware that this exists around them. This chart is included in Chapter 3 and details that workers who are undergoing any form of exertion rapidly become symptomatic when the oxygen level drops below about 19.5%. At 12–16%, they have increased breathing rates, accelerated heartbeat, and impaired attention, thinking, and coordination, even when at rest. As the oxygen concentration continues to drop, workers may experience poor judgement and exhaustion, even with minimal exertion, and they ultimately have heart failure (cardiac arrest) and die.

The following is extracted from the Chemical Safety and Hazard Investigation Board report on the nitrogen fatalities at the Valero Delaware City Refinery.

> 'Workers may be unaware of another dangerous complication: inhaling nitrogen or other inert gas suppresses the brain's breathing reflex response. The breathing reflex is controlled by the amount of carbon dioxide in the blood rather than the shortage of oxygen. Normally, the ability to voluntarily hold one's breath is eventually overwhelmed by the brain's respiratory control centre, which is triggered by the increased carbon dioxide concentration in the blood, combined with a drop in the blood's pH. If high purity nitrogen or other inert gas is inhaled, the body may simply stop breathing, as carbon dioxide accumulation in the blood is insufficient to stimulate the breathing reflex (Lumb, 2005)'.

How long does it take for the oxygen level to drop to a dangerously low level when liquid nitrogen is released into a confined space?

The answer is that when nitrogen is released into the environment, the oxygen level only takes a few minutes to reach deadly concentrations. Also, it is just as important to understand that the victim or victims will not perceive or understand what is happening to them. There is no odour or other indication that they are in danger of being killed.

Case Study Documented by the National Library of Medicine and National Center for Biotechnology Information:

These US National institutions documented the following tragic case study: This case study confirms that the oxygen level will quickly drop to very low levels when nitrogen is released into the environment, especially in confined spaces or areas where there is little air circulation.

> 'A 27-yr-old postgraduate student was found lying on the floor of an unsealed underground dry area, where a valve-opened empty cylinder of liquid nitrogen (150 L) (5.3 Cu Ft) was connected to a cap-removed empty Dewar-flask (10 L) (.3 Cu Ft) via a copper infusion tube.
>
> It is obvious that the student was draining liquid nitrogen from the cylinder into the flask when he was overcome by the lack of oxygen, collapsed and subsequently died. The liquid nitrogen was to be used in a subsequent class study.
>
> No injury was found externally or internally. There were petechiae in the bilateral conjunctive and periorbital skin (pinpoint, round spots forming around the skin around the eyes, sort of like a rash).
>
> The dry area (where the student was collecting the liquid nitrogen), measured 300 × 130 × 260 cm (9.8 Ft. × 4.3 Ft. × 8.5 Ft.)., had a communication to the basement of the research building by a window measuring 90 × 60 cm (3 Ft. × 2 Ft.) in size at 130 cm (4.3 Ft. above the floor).
>
> The scene reconstruction and atmosphere gas analysis revealed that the O_2 concentration at 60 cm (2 Ft.) above the base dropped to 12.0% in 3 min and 10 sec, 10.0% in 8 min and 53 sec, 6.0% in 18 min and 40 sec, and 4.2% in 20 min and 28 sec. The primary cause of death was asphyxia by evaporated liquid nitrogen'.

These extremely low oxygen levels will not sustain life, and this case study confirms that they will occur within minutes of a release of liquid nitrogen into a confined space. Please refer to Figure 1.1 for a visual of the rapid drop in available oxygen resulting in this fatality.

This chart, reconstructed using data from the National Library of Medicine, illustrates the rapid decline in available oxygen that resulted in a fatality. Thanks to the National Library of Medicine for making this data available.

The minimum safe oxygen concentration is 19.5%, as defined by the US Occupational Safety and Health Administration (OSHA). OSHA says health effects start when oxygen concentrations are below 19.5%.

In this incident, the oxygen level dropped below 19.5% in less than a minute and to 12% in 3 minutes. This is well below the level required to sustain life. Within 20 minutes, the oxygen level was at around 4%. This means certain death for everyone who was exposed.

The 27-year-old postgraduate student was found non-responsive in an underground dry area adjacent to a research building lying on plastic pallets. He was lying beside an empty Dewar flask connected to an empty liquid nitrogen cylinder. The valve on the cylinder was in the open position, indicating that the student was filling the flask when he was overcome by the lack of oxygen (due to a release of nitrogen vapours). The larger nitrogen cylinder had been filled the day before this incident. Please refer to Figure 1.2, which illustrates the empty liquid nitrogen cylinder connected to a smaller nitrogen Dewar flask. The student's body was found lying adjacent to the empty cylinder.

Please note the liquid nitrogen cylinder is set up to drain into the Dewar flask, potentially releasing nitrogen vapours into the surrounding environment.

After seeing this, please give more thought to my 'nitrogenated ice cream' story in the Foreword section of this book. Please note from the graph (Figure 1.1) that the available oxygen dropped

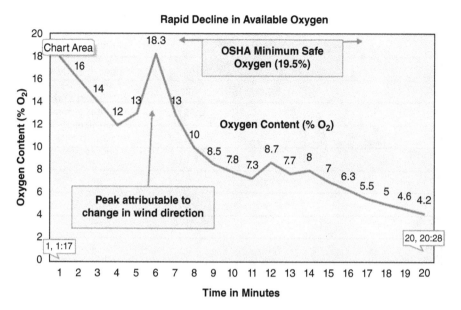

Figure 1.1 National Library of Medicine, National Center for Biotechnology Information data confirm that the oxygen level in a confined space will quickly drop to very low levels in minutes, ensuring certain death for those exposed.

below the OSHA 'Deficient Oxygen' criteria of 19.5% in about 1 minute. In other words, the nitrogen quickly displaced the available oxygen in the room. The available oxygen concentration dropped rapidly and, within the first couple of minutes, was well below that required to sustain human life.

Going back to the ice cream shop, what if the liquid nitrogen release, which occurred overnight the week before, had occurred during the day when the shop was full of teenage attendants and children? Possibly, the alarm would have sounded, and there would be a white fog, but there would have been no odour or other indication that their lives were at risk. The teenage attendant explained that 'they are supposed to go outdoors if the alarm goes off' – she didn't say they will immediately go outdoors if the alarm goes off, and there was no sense of urgency or even an understanding of the potential hazard.

I am very concerned about having liquid nitrogen tanks and fittings inside an enclosure like an ice cream shop – or any enclosure. We also know a liquid nitrogen release occurred at a different ice cream shop in Weston, Florida, sending responders to the hospital for oxygen deprivation injuries. I can't resist thinking if they realized how close they were to death.

The case study, as documented by the National Library of Medicine and National Center for Biotechnology Information, confirms that in the event of a release of liquid nitrogen, the oxygen level would quickly drop to levels well below that required to sustain life.

To expand on this point further, liquid nitrogen has an expansion ratio of 1:696. For example, 10 gallons equals 1.34 cubic feet. If we were to spill only 10 gallons of liquid nitrogen into a small room, for example, an ice cream shop, we would release 933 cubic feet of nitrogen into a confined space.

This would have an effect very similar to the case study shared by the National Library of Medicine, where one person was killed by asphyxiation due to nitrogen, and the Chemical Safety

Figure 1.2 Location of the nitrogen-induced fatality. *Source:* National Library of Medicine (NLN), National Center of Biotechnology Information / Public domain. NLM is part of the National Institutes of Health (NIH), US Department of Health and Human Services, and is located in Bethesda, Maryland.

Board report on the Georgia poultry processing plant, which resulted in six deaths. In this example, the available oxygen level in the room would almost immediately drop to below what is required to support human life.

Transport and storage of liquid nitrogen and other cryogenic gases:
Liquid nitrogen is shipped and stored in specially constructed double-shell tanks. These tanks, sometimes called cylinders, are specially constructed to hold the cryogenic liquids, which are super cold and are designed with a vacuum between the inner and outer shell and several layers of insulation to prevent ambient heat from penetrating the shell to the cold liquid inside. These large cylinders allow the transport of a significant amount of nitrogen; remember, liquid nitrogen expands nearly 700 times when it vapourizes to gaseous nitrogen. Also, liquid nitrogen is a cryogenic liquid meaning that it is supercold and requires very little pressure when compared to nitrogen in a pressurized gaseous state.

Consumers of liquid nitrogen should be aware that ambient heat will penetrate through the cylinder shell and the insulation to the liquid and gradually increase its temperature. This will result in vaporization and a build-up of pressure in the cylinder, referred to as 'head pressure'. This pressure will periodically vent off through the loosely fitting cap or the relief valve. I do not believe liquid nitrogen storage should be in a confined area due to the potential for a release, but if it is, the relief valves must vent outdoors. We must recognize that if the cylinders are located indoors and even with the relief valves venting outdoors, there is still the possibility of a pipe or fitting failure, such as what happened at the ice cream shop.

The uses of nitrogen:

Nitrogen is the most widely used industrial gas, especially by petroleum refineries and petrochemical plants. It is also widely used by pharmaceutical manufacturers, glass and ceramic manufacturers, steel and other metals refining and fabrication companies, and pulp and paper manufacturers. It is commonly known as nitrogen or N_2 but also reported to sometimes be referred to as GAN or GN when in its gaseous form or LIN or LN when in its liquid form (although I have never seen these references). Nitrogen is especially useful to the petroleum refining or petrochemical industry because it is an inert gas that eliminates oxygen, thereby preventing a reaction with hydrocarbons or other chemicals that may result in a fire or explosion.

As a gas, nitrogen may also be used for any or all of the following (Nitrogen has many uses – the following is not intended to be a complete list):
- Nitrogen is used in food processing, purging air conditioning and refrigeration systems, and pressuring aircraft tires. It is also regularly used in petroleum refining and petrochemical processes for tank blanketing, purging oxygen-containing equipment, laboratory use, and instruments and analysers.
- Nitrogen may be supplied by pipeline as compressed gas or delivered in a liquefied state in compressed storage tanks or tank trucks.

In the petroleum refining and petrochemicals industry:
- Preparing equipment by purging hydrocarbons before opening equipment to the atmosphere.
- Remove oxygen from piping and process vessels during start up and before bringing in hydrocarbons.
- Nitrogen is also frequently used as a testing agent when pressure testing the 'tightness' of piping circuits or pressure vessels in preparation for a unit start-up after turnaround or maintenance repairs. This is especially true for tightness testing high-pressure circuits.
- Blanketing tanks to prevent the accumulation of flammable mixtures or products that are sensitive to air.
- For inerting equipment prior to maintenance or mechanical work. For example, ensuring that process vessels such as reactors are essentially free of oxygen for catalyst replacement when the catalyst is pyrophoric. This prevents the catalyst from self-ignition by contacting air.
- Purging equipment of oxygen and other reactive gases before placing it into long-term storage or 'mothballing'. This is done to prevent oxidation or corrosion during storage.
- Pressurized nitrogen may be used to clear plugs in pipelines or to pig lines for cleaning or prior to inspection or maintenance.
- Nitrogen may also be used to pressurize tank cars or railcars to aid in offloading products to storage tanks or other dispositions. Nitrogen is inert and will not react with the hydrocarbons or other chemicals that may be in the railcar.
- For some process and laboratory analyser operations. Process instrumentation or analysers are frequently purged with nitrogen gas.
- For specific welding operations to eliminate air that can interfere with the welding process, contaminating the weld.

In the food processing industry:
- Liquid nitrogen is used as a refrigerant to flash-freeze food products and prepare to ship to the customer.

- Liquid nitrogen is also used as a refrigerant to flash-freeze dairy products to make ice cream or cold treats or fuming drinks such as dragons' breath.

Other uses for nitrogen as a gas:
- Nitrogen is sometimes used to inert aircraft fuel tanks to prevent an explosive mixture in the vapour space. Nitrogen is inert and displaces oxygen, preventing a flammable mixture from occurring in the storage tanks.
- Nitrogen is important to support plant growth and can be added to the soil as an ingredient in fertilizers.
- Nitrogen is frequently used as a filler in light bulbs to eliminate oxygen, thereby preventing combustion of the tungsten filament.
- Nitrogen is widely used in pharmaceuticals and is in every major pharmacological drug class. For example, nitrogen is in most antibiotics.
- Nitrous oxide is sometimes used as an aesthetic.
- Nitrogen forms nitric oxide and nitrogen dioxide with oxygen, ammonia with hydrogen and nitrogen sulphide with sulphur. Some nitrogen compounds can also be formed naturally through biological activity.

As a liquid for storage or transport (liquid nitrogen from pressurized cylinders or tank trucks):
- Liquid nitrogen is a cryogenic liquid, that is, it is stored in a pressure vessel specially designed to maintain liquids at extremely low temperatures. When liquid nitrogen is stored, it is at a temperature of −320 °F (−195 °C), making it very useful as a coolant or refrigerant. Therefore, it has many uses in laboratories, or in the food processing industry, to flash freeze various foods for processing or for shipment. Liquid nitrogen is also used to preserve biological specimens such as sperm, blood or eggs.
- It is also sometimes used in the field to freeze lines when isolation valves are leaking through.
- Liquid nitrogen, when released to near atmospheric pressure, expands 700 times in volume. Therefore, it is also used to store large quantities of nitrogen or transport nitrogen gas in large quantities.

At high temperatures and with the aid of catalysts, nitrogen can combine with some metals to form nitrides. For example, nitrogen and catalysts can combine with lithium, magnesium and titanium at high temperatures to form nitrides. This is a valuable contribution to compounds that can be used as hard coatings or insulators.

Nitrogen is necessary to support some biological processes. For example, it is frequently used as a fertilizer, typically as ammonia or ammonia-based compounds. Some compounds formed with halogens and certain organic compounds can also be explosive.

End of Chapter 1 Review Quiz

1. Why is nitrogen an important compound for use in the petroleum refining and petrochemical process industry?
   Answer(s):

2. Nitrogen is used for petroleum and petrochemical storage tank blanketing to prevent the accumulation of _____ _____ or for products that are sensitive to _____.
   Answer(s):

3 Please explain why liquid nitrogen is frequently used as a refrigerant in food processing and the pharmaceutical industries.
Answer(s):

4 What properties make nitrogen more suitable as a gas to pressurize tank cars or railcars to aid in offloading products to storage tanks or other dispositions?
Answer(s):

5 Why is nitrogen sometimes used to blanket aircraft fuel tanks?
Answer(s):

6 Nitrogen makes up about _____% (by volume) of the air that we breathe.
Answer(s): Please select the correct answer(s).

A 52%

B 85%

C 78%

D 26%

7 Can you name four of the characteristics of nitrogen?
Answer(s):

8 When liquid nitrogen is released into the atmosphere, it quickly forms a white fog by freezing the _____ in the air. It may also freeze anything nearby and can create an oxygen-deficient atmosphere.
Answer(s):

9 One volume of liquid nitrogen expands to approximately _____ volumes of gas.
Answer(s):

10 An unplanned release of liquid nitrogen, for example, from a liquid nitrogen tank, will fill a standard-size room with concentrated nitrogen in _____. People in the room can be quickly overcome by oxygen deprivation, and they can _____.
Answer(s):

11 What mechanism makes nitrogen so dangerous to work with and around?
Answer(s):

12 Cold nitrogen is _____ than air; therefore, it will tend to concentrate along the _____ or when in a building or room, along the _____.
Answer(s):

13 Another significant hazard when working with liquid nitrogen is the possibility of _____ or _____ from contact with the cold liquid.
Answer(s):

14 Why is nitrogen purge used before unit start-up for process piping and vessels?
Answer(s): Please select the correct answer(s).

A As a verification that there are no leaks in the piping or vessels before introducing hydrocarbons.

B To eliminate oxygen from the piping or vessels before introducing hydrocarbons.

C To ensure that the piping or vessels will not fail during the start-up process, releasing hydrocarbons into the atmosphere.

D As a verification that there will be no product contamination during the start-up process.

15 How is nitrogen added to plants to help support plant growth?
Answer(s): Please select the correct answer(s).

A Liquid nitrogen is injected into the soil directly below the plants.

B Nitrogen is converted into ammonia or ammonia-based compounds and used as fertilizer to support plant growth.

C Nitrogen is fed to plants by creating a nitrogen atmosphere during the plant's early life.

D Nitrogen is converted into ammonia, and the ammonia is provided to the plant in the nursery.

16 Liquid nitrogen is a _____ liquid; that is, it is stored in a pressure vessel specially designed to maintain liquids at extremely low temperatures.
Answer(s): Please select the most correct answer(s).

A condensed

B cryogenic

C mixed

D pure

17 When liquid nitrogen is stored, it is at a temperature of _____ °F making it very useful as a coolant or refrigerant.
Answer(s): Please select the most correct answer(s).

A 212 °F

B −200 °F

C 32 °F

D −320 °F

18 Nitrogen is used for tank blanketing to prevent the accumulation of _____ _____ or for products that are sensitive to _____.
Answer(s):

Additional References

Air Products. Safetygram-7, Liquid Nitrogen, 1998. https://www.bnl.gov/esh/shsd/pdf/compressed_gas/safetygram_ln2.pdf

EPA Cameo Nitrogen Safety Datasheet (Nitrogen, Refrigerated Liquid) (Cryogenic Liquid). https://cameochemicals.noaa.gov/chemical/4069

National Library of Medicine and National Center for Biotechnology Information, Evaporated Liquid Nitrogen-Induced Asphyxia: A Case Report. https://www.ncbi.nlm.nih.gov/pmc/articles/PMC2526498/

American National Standard Institute (ANSI) and American Society of Safety Engineers (ASSE), Safety Requirements for Confined Spaces. ANSI/ASSE Z117.1-2003.

American Petroleum Institute (API) API Standard 2217A, 3rd edition, January 2005.

Guidelines for Safe Work in Inert Confined Spaces in the Petroleum and Petrochemical Industries

BP Process Safety Series, Institution of Chemical Engineers (IChemE), Hazards of Nitrogen and Catalyst Handling

Compressed Gas Association, Inc. (CGA), Handbook of Compressed Gases, 4th edition, 1999.

Compressed Gas Association, Inc. (CGA), Accident Prevention in Oxygen-Rich and Oxygen-Deficient Atmospheres. GA P-14-1992.

Compressed Gas Association, Inc. (CGA), CGA, 2001. Safety Bulletin Oxygen-Deficient Atmospheres, SB-2, 4th edition, 2001.

Di Maio VJM, Dana SE, Finkel, and Martin H., 2000. Guidelines for Hot Work in Confined Spaces, Recommended Practices for Industrial Hygienists and Safety Professionals, ASSE, 2000.

Gill JR, Ely SF, and Hua Z. Environmental gas displacement Three accidental deaths in the workplace. *Am J Forensic Med Pathol.* 2002;23:26–30.

Harris, Michael K., Lindsay E. Booher, and Stephanie Carter, 1996 Field Guidelines for Temporary Ventilation of Confined Spaces American Industrial Hygiene Association, 2000.

Kernbach-Wighton G, Kijewski H, Schwanke P, Saur P, Sprung R. Clinical and morphological aspects of death due to liquid nitrogen. *Int J Legal Med.* 1998;111:191–195.

Manwaring, J.C. and C. Conroy, 1990. Occupational Confined Space-Related Fatalities: Surveillance and Prevention *Journal of Safety Research*, Vol. 21, pp. 157–164.

McManus, Neil, 1999. Safety and Health in Confined Spaces, Lewis Publishers/CRC Press.

National Institute of Occupational Safety and Health (NIOSH), 1987. A Guide to Safety in Confined Spaces Publication 87–113, July 1987.

NIOSH, 1986. Preventing Occupational Fatalities in Confined Spaces, DHHS (NIOSH) Publication No. 86–110, January 1986.

OSHA, 1994. Permit-Required Confined Spaces, 29 CFR 1910.146, May 1994.

Rekus, John F., 1994. Complete Confined Spaces Handbook Lewis Publishers/CRC Press.

US Department of Labor Occupational Safety and Health Administration (OSHA). Safety and Health Topics: Confined Spaces www.osha.gov/SLTC/confinedspaces/index.html, June 2006.

US Chemical Safety and Hazard Investigation Board (CSB) Summary Report: Nitrogen Asphyxiation, Union Carbide Corporation, Report No. 1998-05-I-LA.

US Chemical Safety and Hazard Investigation Board (CSB), Safety Bulletin: Hazards of Nitrogen Asphyxiation. No. 2003-10-B, June 2003.

Lion Technology, 10 Hurt in Ice Cream Facility Hazmat Unloading Explosion, Posted on 7/24/2021 by Lauren Scott and Roseanne Bottone. https://www.lion.com/lion-news/july-2021/10-hurt-in-ice-cream-facility-hazmat-unloading-explosion

2

The Properties, Uses, and Safety Hazards of Other Inert Gases

Exposure to nitrogen, especially in confined spaces or working adjacent to where nitrogen is being purged or vented, is the most common cause of asphyxiation. However, nitrogen is not the only gas that can result in severe injury or death due to inhalation exposure. Other gases in this category are argon (Ar), carbon dioxide (CO_2), carbon monoxide (CO), helium (He), neon (Ne), krypton (Kr) and xenon (Xe). These gases are inert and have very similar properties and characteristics to nitrogen. In this chapter, we will discuss each in a little more detail.

Nitrogen and carbon dioxide are inert compound gases that will displace oxygen (O_2) from the environment, especially in confined spaces. Six other elemental gases also fall under this category. They can result in asphyxiation and have caused hypoxia incidents, including death by asphyxiation, in our industry and other industries as well.

These six include argon, helium, krypton, neon, xenon and radon, known as noble gases. They are sometimes called aerogens and are naturally occurring members of group 18 of the periodic table. All but radon are known to have the characteristics of displacing oxygen from the environment and are capable of causing asphyxiation. Therefore, radon is not covered here.

Carbon monoxide is a poisonous gas that functions a little differently in our bodies but nonetheless can result in death or severe injury by interfering with the transport of oxygen to our vital organs.

The inert gases, their primary uses and their properties are listed as follows:

2.1 Argon (Ar)

This is a summary; refer to the Safety Data Sheet for additional information.

Argon is frequently used in the welding of piping and process vessels to create an inert environment, especially when welding alloy metals. Following nitrogen, argon is the most frequently used inert gas in industry and a contributor to asphyxiation incidents, especially where welding is underway. Welders use argon to displace oxygen from the environment to prevent contamination of the weld, especially when welding stainless steel and some other metallurgies. This is typically done by the creation of dams, for example, in the interior of piping and crafting a purge of argon into the area being welded.

> When working with argon, a risk assessment should be conducted and documented in each work area to assess the risks related to the use of argon and select the personal protective equipment that matches the relevant risk. The following recommendations should be considered. It is highly recommended that the welders be protected by the use of either hose

line-supplied breathing air, by the use of a self-contained breathing apparatus (SCBA), or by the establishment of fresh air circulation in the area where they are working. Also, a rescue plan should be established, including the welder equipped with a harness and a retrieval lanyard suitable for immediate rescue in the event they are overcome by the lack of oxygen. For all tasks involving welding with argon in use, there should be an outside observer in direct contact with the welder whose primary responsibility is to alert responders in the event of an emergency. These requirements should all be included in the confined space entry procedures. Please refer to asphyxiation incidents involving argon in the Case Studies section of this book. It is also a good idea that the welders working in a confined space with argon should have a personal oxygen monitor with audible, visual and vibration alarms.

A general description of argon from the US Environmental Protection Agency (EPA) Cameo Chemicals Database and the Safety Data Sheet:
- Argon is a colourless and odourless gas and is totally inert. It does not react with other substances.
- Exposure of the container to prolonged heat or fire can cause it to rupture violently and rocket.
- If liquefied, contact of the very cold liquid with water may cause violent boiling. If the water is hot, a liquid 'superheat' explosion may occur.
- Contact with water in a closed container may cause dangerous pressure to build.

Health hazards of argon from the Argon Safety Data Sheet (SDS):
Although argon is a known asphyxiant, there are no Occupational Safety and Health Administration (OSHA) Permissible Exposure Limit (PEL) or Time Weighted Average (TWA) associated with argon.

The following facts are from the argon SDS:

- Argon is an asphyxiant, and inhalation can result in a lack of oxygen.
- The lack of oxygen can result in death.
- Moderate concentration may cause headache, drowsiness, dizziness, excitation, excess salivation, vomiting and unconsciousness.
- Skin contact: no harm expected.
- Swallowing: this product is a gas at normal temperature and pressure.
- Eye contact: no harm expected.

Precautions for storage and handling argon cylinders:
- Store and use with adequate ventilation.
- Firmly secure cylinders upright to keep them from falling or being knocked over.
- Screw valve protection cap firmly in place by hand.
- Store only where the temperature will not exceed 125 °F (52 °C).
- Store full and empty cylinders separately. Use a first-in, first-out inventory system to prevent storing full cylinders for long periods.
- The cap is intended solely to protect the valve. Never insert an object (e.g., wrench, screwdriver, pry bar) into the cap openings. Doing so may damage the valve and cause a leak.
- Use an adjustable strap wrench to remove over-tight or rusted caps.
- Open the valve slowly. If the valve is hard to open, discontinue use and contact your supplier.
- Never apply flame or localized heat directly to any part of the cylinder. High temperatures may damage the cylinder and could cause the pressure relief device to fail prematurely, venting the cylinder contents.

- Never strike an arc on a compressed gas cylinder or make a cylinder part of an electrical circuit.

Special shipping information:
- Cylinders should be transported in a secure position, in a well-ventilated vehicle.
- Cylinders transported in an enclosed, non-ventilated compartment of a vehicle can present serious safety hazards.

2.2 Carbon Dioxide (CO_2)

This is a summary; refer to the Safety Data Sheet for additional information.

Carbon dioxide is used as a refrigerant and fire extinguishing agent, for example, in fire extinguisher systems installed in compressor or gas turbine generator enclosures or product storage areas where flammables may be stored in containers. Carbon dioxide in the solid state is also known as 'dry ice' and is used in cooling service, food storage, etc. Carbon dioxide is also used to form some rubber and plastics, as an additive to put the fizziness in carbonated beverages, and aiding plants' growth in greenhouses and some nurseries. Carbon dioxide is also known for causing loss of life due to asphyxiation.

The 8-hour Personnel Exposure Limit (TWA) for Carbon dioxide is 5,000 ppm.

A general description of carbon dioxide from the US EPA Cameo Chemicals Database and the Safety Data Sheet:
Carbon dioxide is a colourless, odourless gas at atmospheric temperatures and pressures. Carbon dioxide is colourless. However, at sufficiently high concentrations, it has a sharp, acidic odour. It is relatively non-toxic and non-combustible. It is heavier than air and may asphyxiate by the displacement of air. CO_2 is soluble in water. It can also form carbonic acid, a mild acid. Under prolonged exposure to heat or fire, the container may rupture violently and rocket. It is used frequently to freeze food, to control chemical reactions and as a fire extinguishing agent.

Health hazards of carbon dioxide from the CO_2 Safety Data Sheet:
Contact with liquid may cause cold burns, freezing to the skin or eyes, and frostbite like a burn. Solid can cause cold contact burns. Carbon dioxide causes rapid circulatory insufficiency in high concentrations, even at normal oxygen concentration levels. Symptoms are headache, nausea and vomiting, which may lead to unconsciousness and death.

Inhalation causes increased respiration rate, headache, subtle physiological changes for up to 5% concentration and prolonged exposure. Symptoms may include loss of mobility/consciousness. Victims may not be aware of asphyxiation.

Remove the victim to an uncontaminated area wearing a self-contained breathing apparatus. Keep the victim warm and rested. Call a doctor. Apply artificial respiration if breathing stops.

See Table 2.1: Carbon dioxide effects on the body.

Carbon dioxide and greenhouse gas elimination:
There is a national emphasis on the capture, use and removal of carbon dioxide from the atmosphere as an ongoing environmental initiative in the United States and internationally. This is related to the goal of reduction of greenhouse gases in the environment and their impact on global warming.

This means that more people will be involved in the handling and processing of carbon dioxide in the workplace and, as a result, may be exposed to the hazards associated with this inert gas.

Table 2.1 Overview of carbon dioxide health effects on the human body.

Carbon dioxide concentration (percent)	Time	Effects
2	Several hours	Headache, dyspnea (*dyspnoea*) (difficult or laboured breathing) upon mild exertion
3	1 hour	Mild headache, sweating and dyspnea (*dyspnoea*) (difficult or laboured breathing) when at rest
4–5	Within a few minutes	Headache, dizziness, increased blood pressure, uncomfortable dyspnea (*dyspnoea*) (difficult or laboured breathing)
6	1–2 minutes	Hearing and visual disturbances
	<16 minutes	Headache, dyspnea (*dyspnoea*) (difficult or laboured breathing)
	Several hours	Tremors
7–10	Few minutes	Unconsciousness, near unconsciousness
	1.5 minutes to 1 hour	Headache, increased heart rate, shortness of breath, dizziness, sweating, rapid breathing
17–30	Within 1 minute	Loss of controlled and purposeful activity, unconsciousness, convulsions, coma, death

Source: Extracted from the Environmental Protection Agency, Appendix B.

Chapter 10 is dedicated to additional discussion on Carbon Capture, Use and Storage (CCUS). Please refer to this chapter for additional details on the processes and their associated hazards.

2.3 Carbon Monoxide (CO)

This is a summary; refer to the Safety Data Sheet for additional information.

The following paragraph was from the US Center for Disease Control and Prevention, The National Institute for Occupational Safety and Health (NIOSH):

'Carbon monoxide (CO) is a colorless, odorless and toxic gas predominantly produced by incomplete combustion of carbon-containing materials. Incomplete combustion occurs when insufficient oxygen is used in the fuel (hydrocarbon) burning process. Consequently, more carbon monoxide, in preference to carbon dioxide, is emitted. Some examples of this are the following: vehicle exhausts, fuel-burning furnaces, coal-burning power plants, small gasoline engines, portable gasoline-powered generators, power washers, fireplaces, charcoal grills, marine engines, forklifts, propane-powered heaters, gas water heaters and kerosene heaters. The single largest source of carbon monoxide is vehicle exhaust'.

Exposure to carbon monoxide impedes the blood's ability to carry oxygen to body tissues and vital organs. When carbon monoxide is inhaled, it combines with hemoglobin (*haemoglobin*) (an iron–protein component of red blood cells), producing carboxyhaemoglobin (COHb), which greatly diminishes hemoglobin's (*haemoglobin's*) oxygen-carrying capacity. Hemoglobin's (*Haemoglobin's*) binding affinity for carbon monoxide is 300 times greater than its affinity for oxygen. As a result, small amounts of carbon monoxide can dramatically reduce hemoglobin's (*haemoglobin's*) ability to transport oxygen. Common symptoms of carbon monoxide exposure are headache, nausea, rapid breathing, weakness, exhaustion, dizziness and confusion. Hypoxia (severe oxygen deficiency) due to acute carbon monoxide poisoning may result in reversible neurological effects, or it may result in long-term (and possibly delayed) irreversible neurological (brain damage) or cardiological (heart damage) effects.

Carbon monoxide exposure can be dangerous during pregnancy for both the mother and the developing foetus. Please contact CDC-INFO (800-232-4636) if you have any questions regarding carbon monoxide exposure during pregnancy.

Death can result from exposure to carbon monoxide. In South Louisiana, where I live, we are not immune to hurricanes. We typically see one or two major storms a year affecting some parts of the state. These storms can wreak havoc on the state and its citizens, sometimes causing widespread death and destruction. Unfortunately, the storm doesn't end when the wind stops blowing. People can be severely injured while dealing with the storm aftermath by using chainsaws to cut downed trees, by contacting downed electrical wiring and unfortunately also by exposure to carbon monoxide while running gasoline-powered generators near their homes or sleeping quarters to deal with the electrical power outages that most often accompany these large storms.

Carbon monoxide is used in some industrial settings. It is used in the production of some chemicals, as a reducing agent in metallurgy, and as a fuel, for example, in the manufacture of hydrogen in some industrial processes.

My first real experience with carbon monoxide was when the young daughter of one of my best friends at the time was killed by exposure to carbon monoxide. I'll never forget that experience; she was only 12 years old when this tragic accident happened. She died overnight while sleeping in a travel trailer in a public campground. This tragic incident occurred when a gasoline-powered electrical power generator was used near the trailer where she was sleeping. The carbon monoxide penetrated the trailer, resulting in her death. At this time in our lives, our families frequently camped together. Although we were not with their family on the night of the tragic incident, it hit us all hard and clearly illustrates one of the more significant hazards of carbon monoxide. Gasoline-powered generators should NEVER be placed indoors or even in close proximity to living or sleeping quarters! Carbon monoxide is indeed a silent killer.

The US Occupational Safety and Health Administration (OSHA) issued an OSHA Fact Sheet covering the hazards of carbon monoxide and how it can be identified and avoided. See the Additional Resources Section below for access to the fact sheet. This fact sheet provides the following additional information on carbon monoxide: 'Carbon monoxide (CO) is a poisonous, colorless, odorless and tasteless gas. Although it has no detectable odor, CO is often mixed with other gases that do have an odor. So, you can inhale carbon monoxide along with gases you can smell and not even know that CO is present'.

Carbon monoxide functions a bit differently in the human body than other more typical asphyxiants in that it interferes with the body's ability to transport oxygen to the critical organs, whereas other more common asphyxiants displace oxygen from the environment we breathe. However, the result is the same: our organs are deprived of oxygen and will fail at higher concentrations. The OSHA PEL for CO is 50 parts per million (ppm). OSHA standards prohibit worker exposure to more than 50 parts of CO gas per million parts of air averaged during an 8-hour time period. The National Institute for Occupational Safety and Health (NIOSH) established the immediate danger to life and health (IDLH) at 1,200 ppm.

2.4 Helium (He) (Overview)

Helium is the second-lightest element in the universe (second only to hydrogen), with an atomic number of two. Helium is a noble gas, found in Group 0 on the periodic table, and is plentiful in the atmosphere. It is available at about five parts per million by volume in the atmosphere and is found in abundance in deep seawater or underground sources.

Helium, like nitrogen, is completely odourless and colourless, and it also displaces oxygen similar to nitrogen. This characteristic makes helium particularly dangerous since the victim may be totally unaware of its presence and can easily be overcome due to the lack of oxygen in the breathing space. He's boiling point is extremely low at −452 °F (−269 °C). Its density is also considerably less than air −0.0001785 g/cm^3, making it lighter than other gases like hydrogen and oxygen. Helium is often used as a lifting gas. Additionally, specific heat capacity is quite high compared to other elements as well. So, when used in combination with other substances, it will help absorb more energy from them before reaching a temperature equilibrium. The higher specific heat capacities help stabilize temperatures in their surroundings, and more heat is required to change temperature.

Uses of helium:
Helium is often transported as a pressurized gas or a compressed liquid to save space and efficiency. It is also often used in arc welding as a shielding gas to protect the molten weld pool against elements in the atmosphere, such as nitrogen and hydrogen. When those elements enter the weld pool, they can lead to weld contamination, porosity and cracking.

Helium is also used to trace leaks in refrigeration and other closed systems and as a lifting gas for lighter-than-air aircraft. More recently, helium has been used as a pressure testing (leak test) agent in petroleum refineries and petrochemical plants.

Helium has also evolved as a successful fire suppressant and is used frequently in applications in industrial fire suppression systems. These systems utilize a clean agent, and a total flooding approach, where the agent used is not harmful to the sophisticated electronic equipment, and the total flooding approach displaces the oxygen to stop the fire reaction. Helium may be used as the sole agent, or it may be used in a blend with other inert gases such as argon or nitrogen. The downside of a total flood approach is that extinguishment is caused by the reduction of available oxygen, and this means potential health impacts to personnel.

Helium is also a valuable contributor to many scientific applications. Its extremely low boiling point and thermal conductivity, especially at exceptionally low cryogenic temperatures, help make it an ideal coolant in the laboratory. Since it is from a family of inert gases, it helps prevent the creation of a flammable or explosive environment, protecting people and equipment in laboratory chemical reactions and/or during welding. Liquid helium is frequently used for cooling experiments for specialized electronic equipment and experimental applications.

Helium has also played a major role in the medical field with advances in medical diagnostics such as MRI imaging technology. We believe that helium has helped advance many fields, such as medical imaging and medicines, oil and gas exploration and refining, welding technology, scientific research and many others, and will continue to do so for many decades to come.

Helium characteristics and hazards:
This is a summary; refer to the Safety Data Sheet for additional information.

Note: Helium is considered hazardous by OSHA 29 CFR 1910.1200 (Hazard Communication Standard).

High concentrations may cause asphyxiation:
- Helium is odourless and colourless, and the victim may be unaware of its presence.
- Although helium is a known asphyxiant, there are no OSHA PEL or TWA associated with it.
- Symptoms may include loss of mobility/consciousness. High concentrations can cause passing out and death due to suffocation due to the lack of oxygen. The victim may not be aware of asphyxiation.

- Remove the victim to an uncontaminated area wearing a self-contained breathing apparatus.
- Keep the victim warm and rested.
- Call a doctor.
- Apply artificial respiration if breathing stops.

A general description of helium from the US Environmental Protection Agency (EPA) Cameo Chemicals Database and the Safety Data Sheet:
- Helium is a colourless, odourless and non-combustible gas lighter than air.
- It is sometimes used as a lifting gas.
- Helium can **asphyxiate.**
- Inhalation causes the voice to become squeaky and sounds much like the Mickey Mouse voice.
- Exposure of the container to prolonged heat or fire can cause it to rupture violently and rocket.
- If liquefied, contact of the very cold liquid with water causes violent boiling. Pressures may build to dangerous levels if the liquid contacts water in a closed container.
- Personnel contact with liquid helium will result in frostbite.

Health hazards of helium from the EPA Cameo Chemicals Database:
- Vapours may cause dizziness or asphyxiation without warning, especially when in closed or confined areas.
- Vapours from liquefied gas are initially heavier than air and spread along the ground.
- Contact with gas, liquefied gas or cryogenic liquids may cause burns, severe injury and/or frostbite.
- Helium gas is inert.

Precautions for helium safe handling and storage:
- When moving the helium cylinder, use personal protection equipment.
- Wear leather safety gloves and cover shoes.
- Use a cart or trolley to move the cylinder.
- Store gas in a cool, well-ventilated place. The temperature should not exceed 125° F.
- Firmly strap the cylinder to secure it to a wall or a stable place.
- Close container valves after each use and when the cylinder is empty.

Special shipping information:
- Cylinders should be transported in a secure position, in a well-ventilated vehicle such as the bed of a pickup truck or other open area on a vehicle (never inside a cab or enclosed area with a driver or passengers present).
  - Cylinders transported in an enclosed, non-ventilated compartment of a vehicle can present serious safety hazards.

2.5   Neon (Ne)

This is a summary; refer to the Safety Data Sheet for additional information.

The primary use of neon is to fill lamp bulbs and tubes for interior and exterior lighting, predominately in advertising signs. However, it is also used in high-voltage indicators, lightning arrestors, wave metre tubes and TV tubes. Neon and helium are used in the manufacture of gas lasers. Neon in a liquefied state is also used for some cryogenic refrigerant applications, which do not require a lower temperature than is otherwise available with liquid helium refrigeration.

A general description of neon from the US Environmental Protection Agency (EPA) Cameo Chemicals Database:
- Neon is a colourless, odourless, non-combustible gas.
- It is chemically inert and is non-reactive with other materials.
- The vapours are lighter than air.
- Neon is non-toxic but can act as a simple asphyxiant. The main hazard is the displacement of oxygen from the environment.
- Exposure of the container to prolonged heat or fire may cause it to rupture violently and rocket.

Health hazards of neon from the EPA Cameo Chemicals Database:
- The vapours may cause dizziness or asphyxiation without warning. In an extreme case of asphyxiation, especially in enclosed spaces or areas without adequate ventilation, it can also result in death.
- Vapours from liquefied gas are initially heavier than air and spread along the ground.
- Direct skin contact can cause frostbite.
- High exposure can cause fatigue, vision disturbance, headache, confusion, dizziness and suffocation from lack of oxygen.

2.6 Krypton (Kr)

This is a summary; refer to the Safety Data Sheet for additional information.

The characteristics and uses of krypton:
Krypton is typically used as a gas to fill energy-saving fluorescent lights. It is also sometimes used in small amounts mixed with neon in fluorescent light bulbs.

Krypton may also be used in some specialized lighting for high-speed photography.

A general description of krypton from the US Environmental Protection Agency (EPA) Cameo Chemicals Database:
Krypton is a colourless, odourless compressed gas that occurs naturally in the atmosphere in trace amounts.

It is shipped as a liquid under its vapour pressure. Contact with the liquid may cause frostbite on unprotected skin.

Health hazards of krypton from the EPA Cameo Chemicals Database and the Safety Data Sheet:
- Krypton vapours may cause dizziness or asphyxiation without warning.
- It is inert and can asphyxiate by displacement of air.
- Krypton is about three times heavier than air and may accumulate in low places if a release occurs. This can result in concentrated vapours that can reduce the oxygen content.
  - Vapours from liquefied gas may spread along the ground.
- Exposure of the container to prolonged heat or fire may cause it to rupture violently and rocket.

High concentrations may cause asphyxiation:
- Symptoms may include loss of mobility/consciousness.
- Since the gas is colourless and odourless, the victims may not be aware of asphyxiation.
- Remove the victim to an uncontaminated area wearing a self-contained breathing apparatus.
- Keep the victim warm and rested.
- Call a doctor.
- Apply artificial respiration if breathing stops.

2.7 Xenon (Xe)

This is a summary; refer to the Safety Data Sheet for additional information.

A general description of xenon from the US Environmental Protection Agency (EPA) Cameo Chemicals Database:
Xenon is a colourless and odourless gas and is found in trace quantities in the earth's atmosphere. Xenon is used as a filler gas in flash lamps and arc lamps. It is non-combustible and heavier than air. Xenon may asphyxiate by the displacement of air. Under prolonged exposure to fire or heat, containers may rupture violently and rocket.

Health hazards of xenon from the EPA Cameo Chemicals Database and the Safety Data Sheet:
- Xenon vapours may cause dizziness or asphyxiation without warning.
- It is inert and can asphyxiate by displacement of air.
- Xenon is heavier than air and may accumulate in low places if a release occurs. Vapours from liquefied gas may spread along the ground.
- Exposure of the container to prolonged heat or fire may cause it to rupture violently and rocket.

High concentrations may cause asphyxiation:
- Symptoms may include loss of mobility/consciousness.
- Since the gas is colourless and odourless, the victims may not be aware of asphyxiation.
- Remove the victim to an uncontaminated area wearing a self-contained breathing apparatus.
- Keep the victim warm and rested.
- Call a doctor.
- Apply artificial respiration if breathing stops.

2.8 Light Hydrocarbons

Note: Hydrocarbons are NOT inert and are flammable or explosive. They are included here since some of the effects on personnel are similar.

A general description of propane from the US Safety Data Sheet:
Please refer to the **Safety Data Sheet** for additional information specific to the light hydrocarbon of interest.

Contact with liquefied gas can cause damage (frostbite) due to rapid evaporative cooling. **Suffocation (asphyxiant) hazard – if allowed to accumulate to concentrations that reduce oxygen below safe breathing levels.**

Material can accumulate static charges, which may cause an ignition. Material can release vapours that readily form flammable mixtures. Vapour accumulation could flash and/or explode if ignited.

Light hydrocarbons also represent a significant asphyxiation hazard when vaporizing into the atmosphere or confined space. Saturated hydrocarbon compounds like ethane, butane, propane, and pentane and their saturated cousins, the alkenes and olefins (e.g., ethylene, butylene, propylene, etc.), when vaporized, can become asphyxiants, especially in restricted areas or confined spaces.

Of course, light hydrocarbons are extremely flammable; therefore, we have an incentive to minimize the loss of containment and vaporization of these light hydrocarbons. Once the concentration of hydrocarbons reaches the lower explosive limit, only about 2 or 3% in air for these compounds, an explosion can occur if an ignition source is nearby. Therefore, we generally pay extensive attention to procedures, inspection and maintenance to prevent loss of containment and the vaporization of these compounds.

End of Chapter 2 Review Quiz

1 Several other gases are also asphyxiants and will displace the oxygen in the air we breathe and can kill.

   Name four gases (including nitrogen) that are asphyxiants.

   Answer(s): Choose nitrogen, plus four of the following that are asphyxiants:

A. Nitrogen	**B.** Argon
C. Hexane	**D.** Carbon dioxide
E. Chlorine	**F.** Helium
G. Neon	**H.** Aromatic concentrate
I. Krypton	**J.** Light hydrocarbons like propane or butane
K. Xenon	**L.** Toluene

2 What is the major use of argon in the industry?

   Answer(s):

3 Other asphyxiants include light hydrocarbons, especially in higher concentrations in confined spaces. Please name at least three of the lighter hydrocarbons that share these characteristics.

   Answer(s):

4 What is an evolving use of helium in petroleum refining and petrochemical plants?

   Answer(s): Name as many as you can.

   A As a pressure test medium for piping circuits and process vessels following a major project or turnaround and before unit start-up.

   B As a lifting gas for lighter-than-air balloons to observe a unit start-up in progress.

C. As a sample gas for the newer version of analysers in a process unit analyser shelter.
D. As a cryogenic coolant in nuclear power plant turbine generators.
Answer(s):

5 What sets carbon monoxide apart from other gases known to be asphyxiants?
Answer(s):
Most asphyxiants work by _____ the oxygen from the air we breathe, and we can be killed by oxygen _____ or _____.
Carbon monoxide works by _____ with the transport of oxygen within our body to the internal _____ that depend on oxygen to function.

6 What is common with the classes of light hydrocarbons (methane, ethane, propane, butane, etc.) relative to their potential for causing hypoxia?
Answer(s):

7 Please name as many other gases that you can that have the same consequences as nitrogen when in higher concentration in confined spaces or other areas of poor ventilation.
Answer(s): (Please target to name at least four in this category)
(Correct answers – any four of the following):
- Argon
- Carbon dioxide
- Helium
- Neon
- Xenon
- Krypton
- Light Hydrocarbons (i.e., methane, ethane, propane, butane, pentane and their olefin cousins).

8 Cold nitrogen is _____ than air; therefore, it will tend to concentrate along the _____ or when in a building or room, along the _____.
Answer(s):

9 Another significant hazard when working with liquid nitrogen is the possibility of _____ or _____ from contact with the cold liquid.
Answer(s):

10 What is the OSHA Permissible Exposure Limit or TWA (Time Weighted Average) for exposure to Argon?
Answer(s):
A 5 ppm
B 10 ppm
C 25 ppm
D Although argon is well recognized as an asphyxiant, no permissible exposure limit has been established.

11 What is the primary use of at petroleum refineries and petrochemical plants?
Answer(s):

12 What are the major sources of carbon dioxide in industrial settings?
Answer(s): Answer as many as you can.

13 What is the single largest producer of carbon monoxide?
Answer(s):

14 What is the best resource to find the hazards associated with an inert gas or a toxic chemical?
Answer(s):

15 In this lesson, we learned that light hydrocarbons are very hazardous, including as asphyxiants when in higher concentrations, especially in areas of low ventilation or confined spaces. What remains the primary safety hazard associated with light hydrocarbons?

Additional References

Air Products, Nitrogen safety hazards (and safety hazards of other asphyxiants such as carbon dioxide and argon. Also, safety hazards of oxygen-deficient atmospheres). https://www.airproducts.com/company/sustainability/safetygrams

US Environmental Protection Agency, Appendix B APPENDIX B—Overview of Acute Health Effects (of Carbon Dioxide). https://www.epa.gov/sites/default/files/2015-06/documents/co2appendixb.pdf

Centers for Disease Control and Prevention, The National Institute for Occupational Safety and Health (NIOSH). Welder's Helper Asphyxiated in Argon-inerted Pipe—Alaska. https://www.cdc.gov/niosh/face/stateface/ak/94ak012.html

Dayan A. D., Evaluation of the Human Health Risks of Breathing Atmospheres Containing a Reduced Oxygen Concentration due to the Injection of Argon or Argon + Nitrogen, report to the US EPA, Bad Oldesloe 1994, available from Minimax.

Lambertsen C. J., 1980 Hypoxia, altitude, and acclimatization. Chapter 73 in Medical Physiology ed VB Mountcastle, Vol. 11, 14th ed. CV Mosby, Philadelphia

NFPA, NFPA 2001, 1995 Fall Meeting Report on Proposals, pp. 145 Quincy MA 1995.

Occupational Safety and Health Administration, OSHA Fact Sheet. Carbon Monoxide Poisoning. https://www.osha.gov/sites/default/files/publications/carbonmonoxide-factsheet.pdf

New Jersey Department of Health, Hazardous Substance Fact Sheet – Helium. https://nj.gov/health/eoh/rtkweb/documents/fs/0972.pdf

Propane Safety Data Sheet (SDS), Williams Companies. https://www.williams.com/wp-content/uploads/sites/8/2019/11/propane.pdf

3

The Effects of Nitrogen and Other Asphyxiants on the Body (Oxygen Deprivation)

Air that has become deficient in oxygen can overcome workers in seconds, and this can immediately result in serious injury or even death. In atmospheres of low concentrations of oxygen, for example, where the oxygen has been displaced or diluted by nitrogen, argon, carbon dioxide or other asphyxiants, this can occur with only one breath. Any form of gas that can displace or dilute the oxygen in the air we breathe can have the same result. Oxygen is the only gas humans can breathe and survive. The other significant hazard is that most of these asphyxiants give us absolutely no warning. There is no odour or taste; without an O_2 analyzer, we are clueless that these deadly effects are happening.

Any situation where you can breathe oxygen-deficient air has the same effect. Leakage of nitrogen and other asphyxiants such as argon, helium or carbon dioxide will displace or dilute the oxygen, especially in enclosed areas or confined spaces. For example, areas of poor ventilation may be caused by the construction of temporary enclosures, such as working inside enclosures or tents created by using tarpaulins, inside buildings where nitrogen has been hard piped or inside analyser shelters. If nitrogen is piped into these ventilation-restricted areas and where a minor leak occurs, people can be at risk of asphyxiation. Obviously, the same thing can happen when welding using argon to create an inert atmosphere around the weld to ensure the weld quality is not affected by the oxygen in the atmosphere. In this case, oxygen deprivation can overcome the welders when the argon displaces the oxygen in the area where the welders are working.

Be especially aware when purging process equipment with nitrogen, when opening process equipment in poorly ventilated areas, especially where flanges are at face level, or when using contaminated breathing air. Other at-risk areas include working with liquid nitrogen in pits or close to grade, for example, while using liquid nitrogen for freeze sealing. Cold nitrogen is denser than air, so it will deplete oxygen at floor level and in excavations. You can easily be overcome by leaning down close to a liquid nitrogen spillage, resulting in asphyxiation. More information on the effects of oxygen deprivation is included in Table 3.1.

Welders, when using argon to protect the weld quality, should be protected by the creation of dams using tape or pipe plugs to force the argon into the area where they are welding and to prevent the argon from leaking into the areas where the welders are present. This should always be supplemented by the welders wearing hose line-delivered breathing air, a self-contained breathing apparatus, or forced air ventilation and continuous gas testing to ensure that the atmosphere where the welders are working is safe to support human life (the OSHA requirement is a minimum of at least 19.5% oxygen). I personally prefer a target of a minimum of 21% oxygen (the same as in a normal atmosphere). It is highly recommended that welders use an oxygen detector equipped with audible, visual and vibration alarms. These devices can save the lives of welders who may otherwise be exposed to the hazards of argon in enclosed spaces. More on this will be explored in Chapter 4.

Remember, these same principles apply whenever we are working with any asphyxiant, including nitrogen, argon, carbon dioxide, carbon monoxide, krypton, neon, xenon and light hydrocarbons. Workers' protection should include forced air ventilation or a self-contained breathing apparatus and continuous air quality monitoring to ensure they are protected from being overcome.

Table 3.1 was developed using data from the US Occupational Safety and Health Administration (OSHA) and reflects the effects of nitrogen on the body (hypoxia or oxygen deprivation):

Human beings must breathe oxygen to survive and begin to suffer hypoxia or adverse health effects when the oxygen level of their breathing air drops below 19.5% oxygen. Below 19.5% oxygen, OSHA considers the air to be oxygen-deficient, and the first symptoms of hypoxia may be present.

Table 3.1 The physiological effects of less oxygen (oxygen deprivation).

% Oxygen	Effects expected on the body
23.5	Maximum 'Safe' level (due to the increased risk of fire).[a]
21.0	Typical O_2 concentration in air.
19.5	OSHA considers air below 19.5% to be oxygen-deficient. At this level, workers will start showing signs of oxygen deficiency.
16–19.5	Workers engaged in any form of exertion can rapidly become symptomatic as their tissues fail to obtain the oxygen necessary to function properly (Rom 1992). Increased breathing rates, accelerated heartbeat and impaired thinking or coordination occur more quickly in an oxygen-deficient environment.
	Even a momentary loss of coordination may be devastating to a worker if it occurs while the worker is performing a potentially dangerous activity, such as climbing a ladder.
12–16	Causes tachypnea (increased breathing rates), tachycardia (accelerated heartbeat) and impaired attention, thinking and coordination, even in people who are resting.
	At these low oxygen levels, people are easily confused and can't find themselves out of a confined space that is normally easy to exit.
10–14	Results in faulty judgment, intermittent respiration and exhaustion can be expected even with minimal exertion.
6–10	Results in nausea, vomiting, lethargic movements and perhaps unconsciousness.
<6	Produces convulsions, then apnoea (cessation of breathing), followed by cardiac standstill. These symptoms occur immediately. Even if a worker survives the hypoxic insult, organs may show evidence of hypoxic damage, which may be irreversible; also reported in Rom, W. (see reference in the previous paragraph).

a) The United States' Moon landing program suffered a horrific incident in 1967 when three astronauts (Virgil I. 'Gus' Grissom, Edward H. White and Roger B. Chaffee) were killed in a flash fire in their spacecraft while undergoing a test on the launch pad. Please refer to the reference at the end of this chapter for a report on the investigation into this tragic accident.

The fire was attributed to the combustion of normal on-board fabrics due to the cabin's oxygen-enriched atmosphere. The report indicates that the O_2 content in the cabin was essentially 100% at the time of the fire.

How long does it take for low oxygen concentrations to have an effect?
- When a person enters an oxygen-deprived atmosphere, the oxygen level in the arterial blood drops to a low level within 5–7 seconds.
- Loss of consciousness follows in 10–12 seconds.
- Heart failure and death ensue if the person is not resuscitated within 2–4 minutes.

The facts are that one breath of concentrated nitrogen can result in the death of the victim. The US Chemical Safety and Hazard Investigation Board (CSB) reported in a Nitrogen Safety Bulletin that

> "an atmosphere of only 4 to 6 percent oxygen causes the victim to fall into a coma in less than 40 seconds. Oxygen must be administered within minutes to offer a chance of survival. Even when a victim is rescued and resuscitated, he or she risks cardiac arrest."

Please see the Additional Resources Section for a link to the valuable safety bulletin.

The following quote is from 'Carbon Dioxide as a Fire Suppressant, Examining the Risks', published by the US Environmental Protection Agency (EPA):

> "At concentrations greater than 17 percent, such as those encountered during carbon dioxide fire suppressant use, loss of controlled and purposeful activity, unconsciousness, convulsions, coma and death occurs within 1 minute of initial inhalation of carbon dioxide"

(OSHA 1989, CCOHS 1990, Dalgaard et al. 1972, CATAMA 1953, Lambertsen 1971).

End of Chapter 3 Review Quiz

1. What is the typical oxygen concentration in the air we breathe?

    Answer(s):

    A 23%

    B 40%

    C 12%

    D 19.5%

    E 21%

2. When a worker enters an environment where the oxygen level is significantly reduced, how long is it before the effects take place?

    Answer(s):

    A Seconds

    B Minutes

    C Hours

    D Days

    E Weeks

3. When a person is overcome due to the lack of oxygen, heart failure and death can occur in about _____ to _____ minutes if they are not _____.

    Answer(s):

4 What effects will the workers notice or feel if they enter an oxygen-deprived workspace?
   Answer(s) (Please mark all that apply):
   A Distinct odour or smell
   B Unusual faint feeling
   C Blurred vision
   D No unusual or noticeable effects

5 What is the maximum oxygen concentration allowed by OSHA in a confined space?
   Answer(s):
   A 19.5%
   B 21%
   C 20%
   D 23.5%

6 What is the purpose of OSHA establishing a maximum oxygen concentration? Why a maximum?
   Answer(s):

7 What tactic do welders use to help protect themselves from oxygen deprivation when they are welding using argon to create an inert gas purge around the weld?
   Answer(s):

8 What device is highly recommended for a welder to use when they are using argon inside an enclosed area such as a pipe or vessel?
   Answer(s):

9 Below what concentration of oxygen in breathing air does US OSHA consider the air to be oxygen-deficient?
   Answer(s):
   A 19.5%
   B 20%
   C 21%
   D 23.5%

10 Human beings must breathe oxygen to survive and begin to suffer _____ or adverse _____ effects when the oxygen level of their breathing air drops below _____% oxygen.
   Answer(s):

11 Which two US federal agencies are great resources for additional information on the hazards of nitrogen and oxygen deprivation?
   Answer(s):

Reference

W. Rom., (ed.) (1992). *Environmental and Occupational Medicine*, 2e. Boston: Little, Brown.

Additional References

The US Occupational Safety and Health Administration (OSHA), Standard Interpretations Clarification of OSHA's Requirement for Breathing Air to Have At Least 19.5 Percent Oxygen Content. https://www.osha.gov/laws-regs/standardinterpretations/2007-04-02-0

National Library of Medicine (Liquid Nitrogen Fatality), Evaporated Liquid Nitrogen-induced Asphyxia: A Case Report. https://pubmed.ncbi.nlm.nih.gov/18303222/

Nitrogen Safety Article by the American Institute of Chemical Engineers (AICHE). "Process Safety Beacon: Recent Nitrogen Fatalities Are a Vivid Reminder". https://www.aiche.org/resources/publications/cep/2021/april/process-safety-beacon-recent-nitrogen-fatalities-are-vivid-reminder and https://www.aiche.org/sites/default/files/cep/20210418.pdf

The US Chemical Safety and Hazard Investigation Board "Hazards of Nitrogen Asphyxiation" Issued 2003. https://www.csb.gov/hazards-of-nitrogen-asphyxiation/

The US Chemical Safety and Hazard Investigation Board Nitrogen Asphyxiation Training—"The Hazards of Nitrogen Asphyxiation". https://www.csb.gov/assets/1/20/nitrogen_asphyxiation_bulletin_training_presentation.pdf

The US Chemical Safety and Hazard Investigation Board Nitrogen Safety Bulletin "Hazards of Nitrogen Asphyxiation". https://www.csb.gov/assets/1/20/sb-nitrogen-6-11-031.pdf

The US Chemical Safety and Hazard Investigation Board Nitrogen Safety Bulletin "Nitrogen Enriched Atmospheres Can Kill". https://www.csb.gov/assets/1/20/nitrogen_trifold_22.pdf

The US Department of Health and Human Services, Center for Disease Control and Prevention National Institute of Occupational Safety and Health

Worker Deaths in Confined Spaces—A Summary of NIOSH Surveillance and Investigative Findings. https://www.cdc.gov/niosh/docs/94-103/pdfs/94-103.pdf

The U S Bureau of Labor Statistics Fact Sheet | Fatal Occupational Injuries Involving Confined Spaces | July 2020. https://www.bls.gov/iif/factsheets/fatal-occupational-injuries-confined-spaces-2011-19.htm

The US Environmental Protection Agency (EPA), Carbon Dioxide as a Fire Suppressant: Examining the Risks. https://www.epa.gov/sites/default/files/2015-06/documents/co2report.pdf

Asia Industrial Gases Association, 3 HarbourFront Place, #09-04 HarbourFront Tower 2, Singapore 099254

Hazards of Inert Gases and Oxygen Depletion (AIGA 008/11). https://www.asiaiga.org/uploaded_docs/aiga%20008_11_hazards%20of%20inert%20gases%20and%20oxygen%20depletion.pdf

APOLLO 204 Accident, Report of the Committee on Aeronautical and Space Sciences United States Senate with Additional Views January 30, 1968.

Report of the January 27, 1967, 27.5-second fire, which occurred in the Apollo Command Module during a preflight test on the launch pad. Astronauts Virgil I. Grissom, Edward H. White, and Roger B. Chaffee died in this accident. This is attributable to the excess oxygen concentration in the command module. https://www.nasa.gov/wp-content/uploads/static/history/as204_senate_956.pdf

Wikipedia, Inert gas asphyxiation. https://en.wikipedia.org/wiki/Inert_gas_asphyxiation

US Army, Office of Environment Safety and Health Environment Safety and Health Bulletin

Hazards of Nitrogen Asphyxiation in Confined Spaces. https://safety.army.mil/Portals/0/Documents/ON-DUTY/WORKPLACE/CONFINEDSPACE/Standard/Hazards_of_Nitrogen_Asphyxiation_Dept_of_Energy.pdf

4

Protection for Personnel Against Inert Gas Asphyxiation and/or Cold Burns

Protection for personnel must be planned before conducting any task involving a nitrogen purge of piping, vessels or equipment. This should be covered in detail in the pre-job planning, the job safety analysis and the pre-job safety checklist. Pre-job planning should include which vessels are to be purged, how the purge will be done and specified protection for all personnel during all purge phases.

Finally, the primary way job site hazards are typically managed is through the work permit process. In my Safe Operations Training course, I always say that the work permit is a 'legal and binding agreement between the work permit issuer and the work permit acceptor'. The work permit should clearly spell out the job site hazards and how those hazards will be managed. This means detailing the hazards that the workers may encounter, the precautions required to be exercised by the workers to mitigate the hazards, and any unique or specialized personal protective equipment (PPE) that must be worn or used by the workers.

4.1 Adequate Warning Signs and Barricades

To ensure that all employees working around or near a process vessel or piping circuit that is undergoing nitrogen purge are well aware of the operation and associated hazards, adequate warning signs and barricades should be placed on or near the openings. It is also a good idea to place additional barricades and warning signs at or near the steps or ladders leading to the structure in addition to those located at or near where the nitrogen is being vented. This is to help ensure personnel are aware of the purge in progress and that personnel are not exposed to the nitrogen being purged from the piping or vessel. See Figure 4.1 for an example of a nitrogen purge warning sign, which should be placed near the opening of a process vessel before the inert purge is started and remain in place throughout the purging operation and until the vessel has been either closed or it has been purged with air, and the nitrogen hazard has been eliminated.

This sign and others very similar are available from a wide range of safety suppliers, including safetysign.com, creativesafetysupply.com, compliancesigns.com, Amazon, Walmart and many others.

Unfortunately, this has not always been done in the past. For example, at the Delaware City Refinery (see Case Study Number 2 in Chapter 11 of this book), a sign indicating 'Warning, Confined Space Entry' was placed at the access point to the hydrocracker reactor. However, there was no warning sign to indicate that a nitrogen or inert gas purge was underway in the reactor. See Figure 4.2 for an image of the warning sign that was in place at the time of a tragic incident where

32 *4 Protection for Personnel Against Inert Gas Asphyxiation and/or Cold Burns*

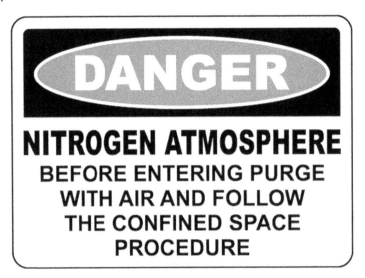

Figure 4.1 Example of a nitrogen inert gas purge warning sign that should be placed near the open access plate of a process vessel undergoing a nitrogen purge. middlenoodle/AdobeStock.

Figure 4.2 Confined space entry sign, which had been placed at the access point to the Valero Delaware City Refinery hydrocracker unit reactor. This is located at the top deck of the hydrocracker unit structure. The sign reads, 'Danger – Confined Space – Do Not Enter Without Permit'. No warning or information was provided to the workers that nitrogen was flowing from the open access point where they were working. *Source:* US Chemical Safety and Hazard Investigation Board (CSB)/Public domain.

two lives were lost when one worker either entered the confined space or fell into the confined space. The second worker followed the first to attempt a rescue and also died in the confined space due to oxygen deprivation.

The Chemical Safety and Hazard Investigation Board (CSB) investigated this incident. It was determined that the first person who entered the reactor was either overcome by a lack of oxygen and fell into the open reactor or possibly intentionally entered the reactor to recover a roll of duct tape. We do know that he was previously observed attempting to recover the roll of duct tape by 'fishing' for it with a piece of wire while positioned directly over the open vessel access point. This task clearly placed the workers at the top reactor access directly in line with where the nitrogen was flowing out. The workers were not protected by hose line-supplied or self-contained breathing air and were directly exposed to the hazard of oxygen deprivation. The CSB investigation found

Figure 4.3 An image of the sign that replaced the 'Danger – Confined Space – Do Not Enter Without a Permit Sign'. *Source:* US Chemical Safety and Hazard Investigation Board (CSB)/Public domain.

that there were no warning signs of nitrogen being used to purge the reactor, and the work permit issued to the contractors indicated nitrogen purge as Not Applicable (NA). No gas test was performed at the reactor outlet flange where the contractors were working.

The second person was observed intentionally entering the reactor in an attempt to rescue the first person, who had already entered or fallen into the reactor and collapsed. We know that in this incident, there was no intent for the workers to enter the confined space; the work permit was issued to install the reactor piping inlet ell to the reactor inlet flange located at the top of the reactor. More is explained on the work permit in the following text.

After the accident and the two fatalities, the Danger – Confined Space sign was replaced with one reading 'DANGER – NITROGEN/INERT GAS PURGE IN PROGRESS. OXYGEN DEFICIENT ATMOSPHERE. DO NOT PASS THIS POINT WITHOUT AUTHORIZATION'. An image of the replacement sign is provided in Figure 4.3.

4.2 Personnel Gas Monitors and Continuous Gas Quality Monitoring and Alarms

It is extremely important that all personnel present in the area near where the nitrogen is venting be equipped with oxygen monitors. Continuous gas detection, including monitoring for oxygen concentration, should be in place for all areas near where nitrogen is venting into the atmosphere. Please see Figure 4.4 for an example of a personal oxygen monitor available from Forensic Detectors and other safety suppliers. Figure 4.5 is an image of a four-gas detector, also available from Forensic Detectors. These gas detectors are available and can be designed for multiple gases, including a wide range of specific gases that may represent unique safety hazards at your facilities.

Similar gas monitors are also readily available for other gases, with the most common monitor being for hydrogen sulphide (H_2S). The most common four-gas personal gas monitor provides continuous testing for oxygen, carbon monoxide (CO), hydrogen sulphide and combustibles (lower

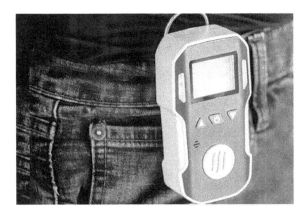

Figure 4.4 A photograph of a personal H$_2$S monitor. *Source:* Courtesy of Forensics Detectors.

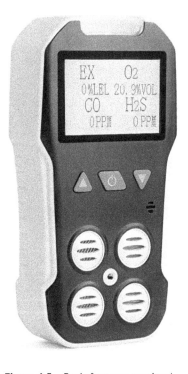

Figure 4.5 Basic four-gas monitor/ USA NIST calibration. *Source:* Courtesy of Forensics Detectors.

explosive limit [LEL]). These monitors typically have audible, visual and vibration alert warnings. I believe that suppliers can meet your needs, whatever they may be, for the gas that represents a potential hazard at your facility.

Continuous gas quality testing with visual and audible alarm capabilities is an alternative to personnel monitors. I supported a construction project on the West Coast of the United States, which involved a very large and deep excavation for a new project foundation that was being built between, and very close to, other large and complex process units. This project was literally sandwiched between an HF alkylation unit, a hydrogen reformer and a large fluid catalytic cracking unit (FCCU).

As part of my consulting services, I worked with the facility, and we contracted for continuous gas monitoring at all four corners at the bottom of the excavation with continuous testing for four of the major concerns of this work being close to these operating areas and substantially below grade. To continually test to ensure continuous safe working conditions for the workers, we were continuously monitoring for oxygen content, hydrogen sulphide, hydrofluoric acid vapours (H-F) and, of course, the percentage of LEL for flammability. All tests were continually being run with alarms programmed for visual and audible indications in the event any trigger points were reached or exceeded for any one of the monitored gases. This continuous testing was done for all periods when workers were inside the excavation and conducted at a very reasonable price for the company. I think this clearly shows that this capability is readily available to industry and should be a routine part of most hazardous work areas where there is the potential for a lack of oxygen or toxic vapours to be present.

Please see Figure 4.4 for an example of a personal gas safety that should be worn by workers in hazardous environments. An oxygen monitor should be worn by workers working near nitrogen purge, and it should be periodically calibrated and tested to ensure that it is maintained in good working order. An H_2S monitor is also extremely valuable and is recommended to be used by all personnel working in areas where the potential for H_2S exists.

Other types of toxic gas or oxygen monitors are available from Forensics Detectors and other safety suppliers. For example, Forensics Detectors provide detectors covering a wide range of toxic chemicals and oxygen deficiency.

As mentioned earlier, other stationary or personal gas detection monitors have multi-gas detection capabilities. Some of these are designed with audible, visual and vibration capabilities to help alert personnel to a lack of oxygen or toxic gas in the area. Figure 4.5 provides an example of a personal four-gas detector.

This four-gas monitor is configured to detect oxygen (O_2), hydrogen sulphide (H_2S), carbon monoxide (CO), and flammables (percentage of LEL), and is available from Forensic Detectors.

4.3 Use of Self-Contained Breathing Apparatus

If personnel are expected to work near vessel openings or open pipe flanges where nitrogen or other inert gases are venting, consideration should be given to requiring the use of a self-contained breathing apparatus for those personnel, with a standby person also in SCBA. See Figure 4.6 for an example of an SCBA.

SCBAs are typically available in negative-pressure or positive-pressure configurations and several sizes. Generally, the higher the tank pressure, the longer the response or rescue breathing time.

I prefer the positive pressure regulator in combination with a full-face mask. The positive pressure will supply air at pressure to the mask anytime the pressure difference is reduced. The pressure inside the mask is slightly above ambient pressure, ensuring positive airflow is always available to the user. This can overcome minor leaks in the mask, preventing the leakage of smoke or vapours into the mask.

The negative pressure system is quickly becoming obsolete; however, it may still be available in some areas. The disadvantage of the negative pressure regulator is that the user must activate

Figure 4.6 Example of a self-contained breathing apparatus (SCBA) with a full-face mask. *Source:* Courtesy of MSA Safety.

the regulator by creating a negative pressure when breathing. In some situations, this may allow outside smoke or vapours into the mask if the mask is leaking.

4.4 The Documented Work Permit (The Authorization for Specified Work to Begin Issued by an Authorized and Designated Individual)

From the US OSHA perspective, the priority features of the work permit are the job site inspection and preplanning aspects of the work permit process. This is clearly spelled out in OSHA Standard 1910.252 (Welding, Cutting and Brazing, General Requirements), where paragraph 1910.252(a)(2)(iv) talks about how the pre-work inspection and authorization for work is by a competent individual ('the individual responsible for authorizing cutting and welding operations'), 'preferably in the form of a written permit'. The work permit is also a specific requirement of 29 CFR 1910.146 Permit-required Confined Entry Standard and 29 CFR 1926.1206 Confined Space Entry during construction activities. This is also true for 29 CFR 1910.119, the Process Safety Management Standard, as it includes specific requirements for Hot Work Permits.

Unfortunately, at the Valero Delaware City Refinery accident site, there was a permit issued for the mechanical task planned for the fateful day (5 November 2005) when two people were asphyxiated due to exposure to a nitrogen environment. According to the CSB Final Report on this incident, To begin work on the hydrocracker unit, a Valero hydrocracker unit operator issued a safe work permit to a Matrix nightshift boilermaker crew to 'install [the] top elbow', or pipe assembly, on R1. The operator told the US Chemical Safety Board (CSB) investigators that he and the Matrix foreman agreed that the crew would only set up the work area and that the foreman would return to the control room after lunch to get a new permit to perform the installation work. However, the permit was not amended to limit the work to "set up only."

The CSB Final Report continued: 'Furthermore, the nitrogen purge status was marked "N/A" on the permit even though the reactor continued to be purged with nitrogen'.

This statement by the CSB indicates that the work permit received by the contractors that fateful evening on 5 November 2005, indicated to them that there was no nitrogen purge on the reactor at that time (Nitrogen Purge N/A or Not Applicable). In hindsight, nitrogen flowed into the reactor and exited at the open access point, the same flange where the contractors would be working and where the roll of duct tape was observed several feet below the open flange. Also, there were no warning signs other than this being a confined space and that entry was not permitted without a confined space permit. There were no warning signs other than the confined space sign and red barricade tape wrapped around the large studs associated with the flange where they would connect the inlet piping.

4.5 The Union Carbide Nitrogen Asphyxiation Incident – Date of Incident 27 March 1998 Investigated by the Chemical Safety and Hazard Investigation Board Final Report Issued 23 February 1999 (Report Number 98-05-I-LA)

The Delaware City Refinery Incident reminds me of another fatal nitrogen asphyxiation incident that occurred at the Union Carbide petrochemical plant in Hahnville, Louisiana, which occurred on 27 March 1998. In that incident, one worker was killed by oxygen deprivation, and another was seriously injured and hospitalized for treatment.

This accident happened when the two workers were inspecting a section of piping, performing a black light inspection at the open end of a 48-inch-wide horizontal section of pipe. The black light inspection was checking for residual organic materials, such as oil or grease, which could react with the oxygen when the line was returned to service. The open end of the large-diameter pipe had been connected to an oxygen feed mixer, which had been disconnected. Both ends of the pipe had been covered with plastic sheeting to keep debris out and to aid in their inspection; the workers added a section of black plastic to block the sunlight. Please see Figure 4.7 for a photograph of one end of the 48-inch diameter pipe with one end covered with clear plastic. In this image, the black plastic can also be seen where it was pulled away after the incident.

The processing unit was shut down, and nitrogen flowed into the pipe and to reactors filled with catalyst. The nitrogen protected the catalyst against moisture during the unit's downtime, which could impact its quality. It also provided an inert blanket for the catalyst. Some of these catalysts are pyrophoric and may self-ignite in the presence of air or oxygen.

For example, in Delaware City, no warning signs were placed near the open section of the pipe to indicate that it was considered a confined space, and no signs or other barricades or warnings were placed to alert the workers about the use of nitrogen. In this case, the workers were also completely unaware of the use of nitrogen in the large-diameter pipe.

I felt the following words from the CSB report are important and are included here. The following is from the Final CSB report on the Union Carbide incident:

'One factor that makes entering a confined space hazardous is that the space may contain a hazardous atmosphere. A confined space may contain a toxic gas, such as hydrogen sulphide, in concentrations hazardous to health. Other confined spaces may contain a nontoxic gas, such as nitrogen, in concentrations that displace the oxygen in the air in the space. The air we normally breathe contains about 21% oxygen, 78% nitrogen and trace amounts of other gases. In this incident, the nitrogen acted as an asphyxiant, causing suffocation by displacing oxygen-containing air'.

It is not necessary for nitrogen to displace all of the 21% of oxygen normally found in the air in order to cause harm to people. OSHA requires that oxygen levels be maintained at or above 19.5% in order to prevent injury to workers. According to the Compressed Gas Association, 'exposure to atmospheres containing 8–10% or less oxygen will bring about unconsciousness without warning

Figure 4.7 Union Carbide Plant; one end of the 48-inch diameter pipe, which was in the process of a black light inspection when the incident occurred. *Source:* Chemical Safety and Hazard Investigation Board (CSB)/Public domain.

and so quickly that the individuals cannot help or protect themselves. Exposure to an atmosphere containing 6–8% oxygen can be fatal in as little as six minutes. Exposure to an atmosphere containing 4–6% oxygen can result in a coma in 40 seconds and subsequent death.

In this incident, no signs were posted at the pipe opening to warn workers and contractors that it was a confined space or that it contained nitrogen. Even if the temporary enclosure had not been erected around the north pipe opening, an employee or a contractor could have been overcome by nitrogen if he or she had merely inserted his or her head into the pipe for a short time. A worker or a contractor could have put his head into the north pipe opening as part of cleaning or inspecting the pipe flange. OSHA has investigated confined space injuries and fatalities at other facilities in which a worker only entered a confined space with his head'.

4.6 Confined Space Entry

No person should be expected to enter any part of an open process vessel with any part of their body unless a confined space entry permit is approved by the authorized personnel. Barricades and warning signs are best placed at the foot of steps or stairs and adjacent to where the venting is occurring, even if the venting occurs up in the structure. The idea is to ensure that personnel do not enter the structure anywhere near the area where the nitrogen is being vented to the atmosphere unless a confined space entry permit is authorized and proper precautions are in place, including personnel oxygen monitors and/or self-contained breathing apparatus.

It is noteworthy that US OSHA in 29 CFR 1910.146 considers entry to have occurred when any part of the body enters the confined space. The following is directly from the regulation; 'Entry includes ensuing work activities in that space and is considered to have occurred as soon as any part of the entrant's body breaks the plane of an opening into the space'.

The sketch in Figure 4.8 illustrates areas around open vessels with nitrogen purges that should have warning signs placed and hard barricades to keep people away who are not involved with the task and without breathing air. Again, normally, the warning signs should be placed at ground level on the stairways or ladder leading to the deck where there are open vessel access plates and at or near where nitrogen is venting. During periods of no entry, it should also be a site requirement to install protective guards over the open manway if workers will be working nearby. Protective guards or lanyards may prevent workers from accidentally falling into the open manway if they are temporarily overcome by oxygen deprivation or toxic gases.

These areas should be provided with warning signs and hard barricades to ensure that only workers properly protected with breathing air are allowed nearby.

The sketch in Figure 4.9 illustrates that we should never (ever!) place our head or any part of our body inside an open access plate unless the vessel has been gas tested and a work permit issued for safe confined space entry.

It only takes one breath of concentrated nitrogen to make you lose consciousness. Before entry with any part of the body, all confined space entry procedures should be in place and approved by authorized personnel.

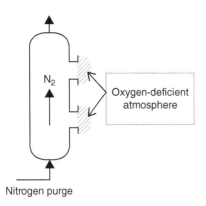

Figure 4.8 Illustration of warning signs near oxygen-deficient atmosphere with nitrogen purge.

4.7 Protection Against Supercold Liquids such as Liquid Nitrogen

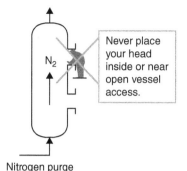

Figure 4.9 Illustration of oxygen-deficient atmosphere. Never place your head or any part of your body near an open vessel access.

Exposure to supercold liquids like liquid nitrogen can subject workers to severe burns due to skin contact with the cold liquid. These burns are no different from a thermal burn in that the skin and flesh can be severely damaged due to exposure to the supercold liquids. As mentioned earlier, liquid nitrogen in storage is around −320 °F (−196 °C) and contact with these extremely low temperatures can result in a severe and very painful burn. When working with these liquids, workers should always protect themselves with proper PPE to protect their bodies, hands and face/eyes from any contact with these supercold liquids. This includes the use of well-insulated or protective aprons or other protective gear, gloves and face and eye protection from cold liquids. See Figure 4.10 for an example of the type of personal protective protection that is readily available and should be used for all potential encounters with these products.

The following is from a Department of the Army Safety Bulletin issued to alert personnel to the hazards of asphyxiation by inert gases. As I have done before, I have taken the liberty of adding several additional items to this list.

This was well written, and I feel it applies to all personnel working with or near any of the family of inert gases. The Additional Resources section of this chapter provides a reference for where this full bulletin can be found.

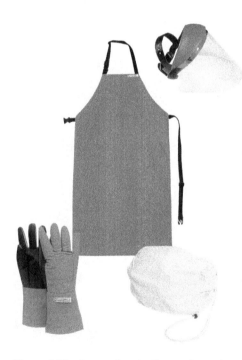

Figure 4.10 Personal protective equipment, including an apron, gloves, gaiters and a face shield, all designed for working around or near supercold liquids such as liquid nitrogen. *Source:* Courtesy of National Safety Apparel.

Management
- In areas where liquid or gaseous nitrogen is used, are there processes for continuously monitoring oxygen levels and alerting workers of oxygen levels less than 19.5%?
- Does our process for work in confined spaces include a provision for placing a trained, equipped worker nearby to observe, communicate with and, if necessary, retrieve overcome workers?
- Do we flow down our safety expectations for working in confined spaces or potential oxygen-deficient atmospheres to subcontractors?
- Do we have positive controls in place to prevent other workers from inadvertently coming into confined or oxygen-deficient spaces?
- Are our existing confined spaces appropriately placarded?

- Do I enforce the full requirements of the Confined Space Entry procedure, including gas testing for lower exposure limit and oxygen before entry?
- Do I always require the use of an oxygen detector when employees are working near inert gas purging or venting operations?

Training
- Are our training programmes for the correct usage of personal protective equipment and ventilating work areas sufficiently comprehensive?
- Does our training programme for confined spaces include a discussion about how temporary confined spaces can be inadvertently established by, for example, hanging a temporary barrier over an opening?
- Is there a provision for annual refresher training for confined space workers?

Individual Worker
- Do I understand what constitutes a confined space, and do I know how to work safely in one?
- Do I always use appropriate PPE (self-contained breathing apparatus and backup breathing air supplies) when working around nitrogen gas?
- Do I always use the 'buddy system' when entering a confined space?
- Do I know to stop work if I see a potentially dangerous situation?
- Do I understand the procedures for worker retrieval should a worker lose consciousness while working in a confined space?
- Do I fully understand that I should always call for emergency responders and never enter a confined space to assist others unless I have been trained and have the appropriate rescue gear (including a self-contained breathing apparatus)?

End of Chapter 4 Review Quiz

1. At what point in a nitrogen purge procedure is the planning for personnel protection considered?

   Answer(s) (select all that apply):

   A During the pre-job planning.

   B During the job safety analysis.

   C During the pre-job safety checklist.

   D Before any work starts in the field.

   E After the purge is underway.

2. What are examples of ways personnel are protected against the hazard of oxygen deprivation?

   Answer(s) (select all that apply):

   A Safety warning signs indicating nitrogen purge in progress.

   B Personnel-restricted entry barricades.

   C Personal oxygen monitors with alarms.

   D Continuous gas testing with audible and visual alarms.

   E A self-contained breathing apparatus (SCBA) is required.

3 Which US OSHA standard covers the hazards and requirements associated with personnel entering into a confined space?

Answer(s):

4 What are the two main personnel safety hazards associated with liquid nitrogen?

Answer(s):

5 How can cold burns be prevented when working around or with liquid nitrogen?

Answer(s):

6 At what temperature is liquid nitrogen when it is being stored in a cryogenic storage tank?

Answer(s):

7 What does US OSHA define as entry into a process vessel?

Answer(s) (please select the most correct answer):

A OSHA considers entry any time the head is placed inside a confined space.

B OSHA considers entry to have occurred anytime the hands or feet break the plane of the process vessel, which is considered inside the vessel.

C OSHA considers entry to occur anytime work is done inside the vessel and the person is inside the space.

D OSHA considers entry as ensuing work activities in that space and is considered to have occurred as soon as any part of the entrant's body breaks the plane of an opening into the space.

8 When a nitrogen purge is underway, where is the best place to place the warning signs and personnel restriction barricades?

Answer(s):

9 It only takes one breath of _____ nitrogen to make you lose consciousness.

Answer(s):

10 What steps should a worker take if they observe a team member collapse inside a confined space?

Answer(s):

11 On 5 November 2005, at the Valero Delaware City Refinery, the workers were tasked with installing the top reactor entry piping on the vessel manway. The vessel was under a nitrogen purge to protect the catalyst. What was missing that should have provided the workers with warnings or key information to help protect the workers?

Answer(s) (select all that apply):

A Safety warning signs indicating nitrogen purge in progress.

B Personnel-restricted entry barricades.

C Personal oxygen monitors with alarms.

D Continuous gas testing with audible and visual alarms.

E Required use of a self-contained breathing apparatus (SCBA).

F An accurate work permit indicating nitrogen purge in progress.

G Protective entry guards or lanyards to prevent workers from falling into the open reactor.

12 What is the significant difference between a positive-pressure SCBA and a negative-pressure SCBA?

Answer(s):

13 US OSHA in 29 CFR 1910.146 considers entry to have occurred when _____ part of the body enters the confined space.

Answer(s):

14 The following lists several potential hazards that may be present when entering a confined space.

Answer(s) (select all that apply):

A The space may be oxygen deficient.

B The space may contain toxic hazards such as hydrogen sulphide or benzene.

C The space may contain engulfment hazards such as a catalyst.

D The space may contain hydrocarbons that are within the flammable range.

E None of the above.

F All of the above.

Additional References

Chemical Safety and Hazard Investigation Board (CSB), Case Study: Confined Space Entry—Worker and Would-be Rescuer Asphyxiated (Incident date: November 5, 2005). No. 2006-02-I-DE Published November 2, 2006.

Self-contained breathing apparatus: https://en.wikipedia.org/wiki/Self-contained_breathing_apparatus#:~:text=An%20open%20circuit%20SCBA%20typically,and%20mounted%20on%20a%20framed

US Chemical Safety and Hazard Investigation Board Final Incident Report – "Valero Delaware City Refinery Asphyxiation Incident". https://www.csb.gov/valero-delaware-city-refinery-asphyxiation-incident/

British Health and Safety Executive Permit to Work Systems. https://www.hse.gov.uk/comah/sragtech/techmeaspermit.htm

The Chemical Safety and Hazard Investigation Board Union Carbide Summary Report (Incident date: March 27, 1998).

Nitrogen Asphyxiation (1 Death, 1 Injury). https://www.csb.gov/assets/1/20/final_union_carbide_report.pdf

Compressed Air Best Practices. https://www.airbestpractices.com/
Compressed Gas Association; Safety Bulletin SB-2 (1992).
US Chemical Safety Board and Hazard Investigation Board (SCSB), Nitrogen Safety Bulletin. https://www.csb.gov/assets/1/20/sb-nitrogen-6-11-03.pdf?13883
Air Liquide, Nitrogen Safety Data Sheet. https://ca.healthcare.airliquide.com/sites/alh_ca/files/2022-07/alh_nitrogen_compressed_7727-37-9_ca-1001-00181_en.pdf
National Institutes of Health, Protocol for Use and Maintenance of Oxygen Monitoring Devices 2022. Authored by the Division of Occupational Health and Safety (DOHS).
Oxygen Monitoring Program Manager. https://ors.od.nih.gov/sr/dohs/Documents/protocol-for-use-and-maintenance-of-oxygen-monitoring-devices.pdf
Compressed Gas Association (CGA) P-12-2017 Safe Handling of Cryogenic Liquids. https://portal.cganet.com/Publication/Details.aspx?id=P-12
British Compressed Gas Association (BCGA). The Safe Use of Liquid Nitrogen Dewars – Code of Practice 30. www.bcga.co.uk
Lawrence Berkeley National Lab Safety Manual. Chapter 29 Safe Handling of Cryogenic Liquids – Safety Manual. https://commons.lbl.gov/display/rpm2/Cryogenic+Liquid+Hazards+and+Controls
Mine Safety Appliance Company (MSA). Monitoring for Oxygen Deficiency: MRI Units – Data Sheet (March 2016). http://s7d9.scene7.com/is/content/minesafetyappliances/Toxgard%20II%20Instruction%20Manual%20-%20EN
Mine Safety Appliance Company (MSA). Toxgard® II Monitor – Instruction Manual, Rev. 9. http://s7d9.scene7.com/is/content/minesafetyappliances/Toxgard%20II%20Instruction%20Manual%20-%20EN
National Fire Protection Association (NFPA). Standard 45-2019. Standard on Fire Protection for Laboratories Using Chemicals.
https://www.nfpa.org/codes-and-standards/all-codes-and-standards/list-of-codes-and-standards/detail?code=45
National Fire Protection Association (NFPA). Standard 55-2020. Compressed Gases and
Cryogenic Fluids Code. https://www.nfpa.org/codes-and-standards/all-codes-and-standards/list-of-codes-and-standards/detail?code=55
NIH Design Requirements Manual (DRM). Section 13.10.7 Liquid Nitrogen (2016). https://orf.od.nih.gov/TechnicalResources/Pages/DesignRequirementsManual2016.aspx
NIH ORF Occupational Health & Safety Manual, Section 3-3 – Confined Space. https://orf.od.nih.gov/TechnicalResources/Pages/DesignRequirementsManual2016.aspx
Department of the Army Office of Environment, Safety, and Health (Safety Bulletin 2005-17, December 2005), Hazards of Nitrogen Asphyxiation in Confined Spaces. https://safety.army.mil/Portals/0/Documents/ON-DUTY/WORKPLACE/CONFINEDSPACE/Standard/Hazards_of_Nitrogen_Asphyxiation_Dept_of_Energy.pdf

Safety Suppliers (Reference only)

- Forensics Detectors.
  https://www.forensicsdetectors.com/
- Industrial Scientific Corporation.
  https://www.indsci.com/en/

- Industrial Safety Products.
  https://www.industrialsafetyproducts.com/
- Air Systems International.
  https://www.airsystems.com/
- Western Safety Products.
  https://westernsafety.com/
- Empire Safety and Supply.
  https://www.empiresafety.com/

5

Confined Space Entry – The Occupational Safety and Health Administration Standard (29 CFR 1910.146) and Some Key OSHA 'Letters of Interpretations'

The US Occupational Safety and Health Administration (OSHA) recognized the hazards of entering and working in confined spaces and promulgated the regulation on 14 January 1993, shortly after the Process Safety Standard (29 CFR 1910.119) was recorded in the Federal Register (1 June 1992).

They developed a detailed and comprehensive regulation laying out requirements for preparing to enter and when entering a confined space. The regulation is provided here and is readily available online. I recommend always checking the online version to ensure you are accessing the most current version. The online OSHA version is available at the following link: https://www.osha.gov/laws-regs/regulations/standardnumber/1910/1910.146

OSHA also routinely responds to questions or clarification about the regulation requirements through periodic interpretations of the meaning of the regulation and its requirements. This chapter provides the regulation as it exists at the time of this writing and several of the most important interpretations associated with the permit-required confined space entry regulation. You may find your answer here if you have questions for OSHA regarding Confined Space Entry. However, to be sure you have the latest information, I encourage you to contact OSHA directly.

5.1 The US OSHA Confined Space Regulation '29 CFR 1910.146 – Permit Required Confined Spaces'

Part Number: 1910

Part Number Title: Occupational Safety and Health Standards

Subpart: 1910 Subpart J

Subpart Title: General Environmental Controls

Standard Number: 1910.146

Title: Permit-required confined spaces

Appendix: A, B, C, D, E, F

GPO Source: e-CFR

1910.146(a)

Scope and application. This section contains requirements for practices and procedures to protect employees in general industry from the hazards of entry into permit-required confined spaces. This section does not apply to agriculture, to construction, or to shipyard employment (parts 1928, 1926, and 1915 of this chapter, respectively).

1910.146(b)

Definitions.

Acceptable entry conditions means the conditions that must exist in a permit space to allow entry and to ensure that employees involved with a permit-required confined space entry can safely enter into and work within the space.

Attendant means an individual stationed outside one or more permit spaces who monitors the authorized entrants and who performs all attendant's duties assigned in the employer's permit space program.

Authorized entrant means an employee who is authorized by the employer to enter a permit space.

Blanking or blinding means the absolute closure of a pipe, line, or duct by the fastening of a solid plate (such as a spectacle blind or a skillet blind) that completely covers the bore and that is capable of withstanding the maximum pressure of the pipe, line, or duct with no leakage beyond the plate.

Confined space means a space that:

(1) Is large enough and so configured that an employee can bodily enter and perform assigned work; and

(2) Has limited or restricted means for entry or exit (for example, tanks, vessels, silos, storage bins, hoppers, vaults, and pits are spaces that may have limited means of entry.); and

(3) Is not designed for continuous employee occupancy.

Double block and bleed means the closure of a line, duct, or pipe by closing and locking or tagging two in-line valves and by opening and locking or tagging a drain or vent valve in the line between the two closed valves.

Emergency means any occurrence (including any failure of hazard control or monitoring equipment) or event internal or external to the permit space that could endanger entrants.

Engulfment means the surrounding and effective capture of a person by a liquid or finely divided (flowable) solid substance that can be aspirated to cause death by filling or plugging the respiratory system or that can exert enough force on the body to cause death by strangulation, constriction, or crushing.

Entry means the action by which a person passes through an opening into a permit-required confined space. Entry includes ensuing work activities in that space and is considered to have occurred as soon as any part of the entrant's body breaks the plane of an opening into the space.

Entry permit (permit) means the written or printed document that is provided by the employer to allow and control entry into a permit space and that contains the information specified in paragraph (f) of this section.

Entry supervisor means the person (such as the employer, foreman, or crew chief) responsible for determining if acceptable entry conditions are present at a permit space where entry is planned, for authorizing entry and overseeing entry operations, and for terminating entry as required by this section.

Note: An entry supervisor also may serve as an attendant or as an authorized entrant, as long as that person is trained and equipped as required by this section for each role he or she fills. Also, the duties of entry supervisor may be passed from one individual to another during the course of an entry operation.

Hazardous atmosphere means an atmosphere that may expose employees to the risk of death, incapacitation, impairment of ability to self-rescue (that is, escape unaided from a permit space), injury, or acute illness from one or more of the following causes:

(1) Flammable gas, vapor, or mist in excess of 10 percent of its lower flammable limit (LFL);

(2) Airborne combustible dust at a concentration that meets or exceeds its LFL;

Note: This concentration may be approximated as a condition in which the dust obscures vision at a distance of 5 feet (1.52 m) or less.

(3) Atmospheric oxygen concentration below 19.5 percent or above 23.5 percent;

(4) Atmospheric concentration of any substance for which a dose or a permissible exposure limit is published in subpart G, Occupational Health and Environmental Control, or in subpart Z, Toxic and Hazardous Substances, of this part and which could result in employee exposure in excess of its dose or permissible exposure limit;

Note: An atmospheric concentration of any substance that is not capable of causing death, incapacitation, impairment of ability to self-rescue, injury, or acute illness due to its health effects is not covered by this provision.

(5) Any other atmospheric condition that is immediately dangerous to life or health.

Note: For air contaminants for which OSHA has not determined a dose or permissible exposure limit, other sources of information, such as Material Safety Data Sheets that comply with the Hazard Communication Standard, 1910.1200 of this part, published information, and internal documents can provide guidance in establishing acceptable atmospheric conditions.

Hot work permit means the employer's written authorization to perform operations (for example, riveting, welding, cutting, burning, and heating) capable of providing a source of ignition.

Immediately dangerous to life or health (IDLH) means any condition that poses an immediate or delayed threat to life or that would cause irreversible adverse health effects or that would interfere with an individual's ability to escape unaided from a permit space.

Note: Some materials - hydrogen fluoride gas and cadmium vapor, for example - may produce immediate transient effects that, even if severe, may pass without medical attention, but are followed by sudden, possibly fatal collapse 12-72 hours after exposure. The victim "feels normal" from recovery from transient effects until collapse. Such materials in hazardous quantities are considered to be "immediately" dangerous to life or health.

Inerting means the displacement of the atmosphere in a permit space by a noncombustible gas (such as nitrogen) to such an extent that the resulting atmosphere is noncombustible.

Note: This procedure produces an IDLH oxygen-deficient atmosphere.

Isolation means the process by which a permit space is removed from service and completely protected against the release of energy and material into the space by such means as: blanking or blinding; misaligning or removing sections of lines, pipes, or ducts; a double block and bleed system; lockout or tagout of all sources of energy; or blocking or disconnecting all mechanical linkages.

Line breaking means the intentional opening of a pipe, line, or duct that is or has been carrying flammable, corrosive, or toxic material, an inert gas, or any fluid at a volume, pressure, or temperature capable of causing injury.

Non-permit confined space means a confined space that does not contain or, with respect to atmospheric hazards, have the potential to contain any hazard capable of causing death or serious physical harm.

Oxygen deficient atmosphere means an atmosphere containing less than 19.5 percent oxygen by volume.

Oxygen enriched atmosphere means an atmosphere containing more than 23.5 percent oxygen by volume.

Permit-required confined space (permit space) means a confined space that has one or more of the following characteristics:

(1) Contains or has a potential to contain a hazardous atmosphere;

(2) Contains a material that has the potential for engulfing an entrant;

(3) Has an internal configuration such that an entrant could be trapped or asphyxiated by inwardly converging walls or by a floor which slopes downward and tapers to a smaller cross-section; or

(4) Contains any other recognized serious safety or health hazard.

Permit-required confined space program (permit space program) means the employer's overall program for controlling, and, where appropriate, for protecting employees from, permit space hazards and for regulating employee entry into permit spaces.

Permit system means the employer's written procedure for preparing and issuing permits for entry and for returning the permit space to service following termination of entry.

Prohibited condition means any condition in a permit space that is not allowed by the permit during the period when entry is authorized.

Rescue service means the personnel designated to rescue employees from permit spaces.

Retrieval system means the equipment (including a retrieval line, chest or full-body harness, wristlets, if appropriate, and a lifting device or anchor) used for non-entry rescue of persons from permit spaces.

Testing means the process by which the hazards that may confront entrants of a permit space are identified and evaluated. Testing includes specifying the tests that are to be performed in the permit space.

Note: Testing enables employers both to devise and implement adequate control measures for the protection of authorized entrants and to determine if acceptable entry conditions are present immediately prior to, and during, entry.

1910.146(c)

General requirements.

1910.146(c)(1)

The employer shall evaluate the workplace to determine if any spaces are permit-required confined spaces.

Note: Proper application of the decision flow chart in appendix A to 1910.146 would facilitate compliance with this requirement.

1910.146(c)(2)

If the workplace contains permit spaces, the employer shall inform exposed employees, by posting danger signs or by any other equally effective means, of the existence and location of and the danger posed by the permit spaces.

Note: A sign reading "DANGER - PERMIT-REQUIRED CONFINED SPACE, DO NOT ENTER" or using other similar language would satisfy the requirement for a sign.

1910.146(c)(3)

If the employer decides that its employees will not enter permit spaces, the employer shall take effective measures to prevent its employees from entering the permit spaces and shall comply with paragraphs (c)(1), (c)(2), (c)(6), and (c)(8) of this section.

1910.146(c)(4)

If the employer decides that its employees will enter permit spaces, the employer shall develop and implement a written permit space program that complies with this section. The written program shall be available for inspection by employees and their authorized representatives.

1910.146(c)(5)

An employer may use the alternate procedures specified in paragraph (c)(5)(ii) of this section for entering a permit space under the conditions set forth in paragraph (c)(5)(i) of this section.

1910.146(c)(5)(i)

An employer whose employees enter a permit space need not comply with paragraphs (d) through (f) and (h) through (k) of this section, provided that:

1910.146(c)(5)(i)(A)

The employer can demonstrate that the only hazard posed by the permit space is an actual or potential hazardous atmosphere;

1910.146(c)(5)(i)(B)

The employer can demonstrate that continuous forced air ventilation alone is sufficient to maintain that permit space safe for entry;

1910.146(c)(5)(i)(C)

The employer develops monitoring and inspection data that supports the demonstrations required by paragraphs (c)(5)(i)(A) and (c)(5)(i)(B) of this section;

1910.146(c)(5)(i)(D)

If an initial entry of the permit space is necessary to obtain the data required by paragraph (c)(5)(i)(C) of this section, the entry is performed in compliance with paragraphs (d) through (k) of this section;

1910.146(c)(5)(i)(E)

The determinations and supporting data required by paragraphs (c)(5)(i)(A), (c)(5)(i)(B), and (c)(5)(i)(C) of this section are documented by the employer and are made available to each employee who enters the permit space under the terms of paragraph (c)(5) of this section or to that employee's authorized representative; and

1910.146(c)(5)(i)(F)

Entry into the permit space under the terms of paragraph (c)(5)(i) of this section is performed in accordance with the requirements of paragraph (c)(5)(ii) of this section.

Note: See paragraph (c)(7) of this section for reclassification of a permit space after all hazards within the space have been eliminated.

1910.146(c)(5)(ii)

The following requirements apply to entry into permit spaces that meet the conditions set forth in paragraph (c)(5)(i) of this section.

1910.146(c)(5)(ii)(A)

Any conditions making it unsafe to remove an entrance cover shall be eliminated before the cover is removed.

1910.146(c)(5)(ii)(B)

When entrance covers are removed, the opening shall be promptly guarded by a railing, temporary cover, or other temporary barrier that will prevent an accidental fall through the opening and that will protect each employee working in the space from foreign objects entering the space.

1910.146(c)(5)(ii)(C)

Before an employee enters the space, the internal atmosphere shall be tested, with a calibrated direct-reading instrument, for oxygen content, for flammable gases and vapors, and for potential toxic air contaminants, in that order. Any employee who enters the space, or that employee's authorized representative, shall be provided an opportunity to observe the pre-entry testing required by this paragraph.

1910.146(c)(5)(ii)(D)

There may be no hazardous atmosphere within the space whenever any employee is inside the space.

1910.146(c)(5)(ii)(E)

Continuous forced air ventilation shall be used, as follows:

1910.146(c)(5)(ii)(E)(1)

An employee may not enter the space until the forced air ventilation has eliminated any hazardous atmosphere;

1910.146(c)(5)(ii)(E)(2)

The forced air ventilation shall be so directed as to ventilate the immediate areas where an employee is or will be present within the space and shall continue until all employees have left the space;

1910.146(c)(5)(ii)(E)(3)

The air supply for the forced air ventilation shall be from a clean source and may not increase the hazards in the space.

1910.146(c)(5)(ii)(F)

The atmosphere within the space shall be periodically tested as necessary to ensure that the continuous forced air ventilation is preventing the accumulation of a hazardous atmosphere. Any employee who enters the space, or that employee's authorized representative, shall be provided with an opportunity to observe the periodic testing required by this paragraph.

1910.146(c)(5)(ii)(G)

If a hazardous atmosphere is detected during entry:

1910.146(c)(5)(ii)(G)(1)

Each employee shall leave the space immediately;

1910.146(c)(5)(ii)(G)(2)

The space shall be evaluated to determine how the hazardous atmosphere developed; and

1910.146(c)(5)(ii)(G)(3)

Measures shall be implemented to protect employees from the hazardous atmosphere before any subsequent entry takes place.

1910.146(c)(5)(ii)(H)

The employer shall verify that the space is safe for entry and that the pre-entry measures required by paragraph (c)(5)(ii) of this section have been taken, through a written certification that contains the date, the location of the space, and the signature of the person providing the certification. The certification shall be made before entry and shall be made available to each employee entering the space or to that employee's authorized representative.

1910.146(c)(6)

When there are changes in the use or configuration of a non-permit confined space that might increase the hazards to entrants, the employer shall reevaluate that space and, if necessary, reclassify it as a permit-required confined space.

1910.146(c)(7)

A space classified by the employer as a permit-required confined space may be reclassified as a non-permit confined space under the following procedures:

1910.146(c)(7)(i)

If the permit space poses no actual or potential atmospheric hazards and if all hazards within the space are eliminated without entry into the space, the permit space may be reclassified as a non-permit confined space for as long as the non-atmospheric hazards remain eliminated.

1910.146(c)(7)(ii)

If it is necessary to enter the permit space to eliminate hazards, such entry shall be performed under paragraphs (d) through (k) of this section. If testing and inspection during that entry demonstrate that the hazards within the permit space have been eliminated, the permit space may be reclassified as a non-permit confined space for as long as the hazards remain eliminated.

Note: Control of atmospheric hazards through forced air ventilation does not constitute elimination of the hazards. Paragraph (c)(5) covers permit space entry where the employer can demonstrate that forced air ventilation alone will control all hazards in the space.

1910.146(c)(7)(iii)

The employer shall document the basis for determining that all hazards in a permit space have been eliminated, through a certification that contains the date, the location of the space, and the signature of the person making the determination. The certification shall be made available to each employee entering the space or to that employee's authorized representative.

1910.146(c)(7)(iv)

If hazards arise within a permit space that has been declassified to a non-permit space under paragraph (c)(7) of this section, each employee in the space shall exit the space. The employer shall then reevaluate the space and determine whether it must be reclassified as a permit space, in accordance with other applicable provisions of this section.

1910.146(c)(8)

When an employer (host employer) arranges to have employees of another employer (contractor) perform work that involves permit space entry, the host employer shall:

1910.146(c)(8)(i)

Inform the contractor that the workplace contains permit spaces and that permit space entry is allowed only through compliance with a permit space program meeting the requirements of this section;

1910.146(c)(8)(ii)

Apprise the contractor of the elements, including the hazards identified and the host employer's experience with the space, that make the space in question a permit space;

1910.146(c)(8)(iii)

Apprise the contractor of any precautions or procedures that the host employer has implemented for the protection of employees in or near permit spaces where contractor personnel will be working;

1910.146(c)(8)(iv)

Coordinate entry operations with the contractor, when both host employer personnel and contractor personnel will be working in or near permit spaces, as required by paragraph (d)(11) of this section; and

1910.146(c)(8)(v)

Debrief the contractor at the conclusion of the entry operations regarding the permit space program followed and regarding any hazards confronted or created in permit spaces during entry operations.

1910.146(c)(9)

In addition to complying with the permit space requirements that apply to all employers, each contractor who is retained to perform permit space entry operations shall:

1910.146(c)(9)(i)

Obtain any available information regarding permit space hazards and entry operations from the host employer;

1910.146(c)(9)(ii)

Coordinate entry operations with the host employer, when both host employer personnel and contractor personnel will be working in or near permit spaces, as required by paragraph (d)(11) of this section; and

1910.146(c)(9)(iii)

Inform the host employer of the permit space program that the contractor will follow and of any hazards confronted or created in permit spaces, either through a debriefing or during the entry operation.

1910.146(d)

Permit-required confined space program (permit space program). Under the permit space program required by paragraph (c)(4) of this section, the employer shall:

1910.146(d)(1)

Implement the measures necessary to prevent unauthorized entry;

1910.146(d)(2)

Identify and evaluate the hazards of permit spaces before employees enter them;

1910.146(d)(3)

Develop and implement the means, procedures, and practices necessary for safe permit space entry operations, including, but not limited to, the following:

1910.146(d)(3)(i)

Specifying acceptable entry conditions;

1910.146(d)(3)(ii)

Providing each authorized entrant or that employee's authorized representative with the opportunity to observe any monitoring or testing of permit spaces;

1910.146(d)(3)(iii)

Isolating the permit space;

1910.146(d)(3)(iv)

Purging, inerting, flushing, or ventilating the permit space as necessary to eliminate or control atmospheric hazards;

1910.146(d)(3)(v)

Providing pedestrian, vehicle, or other barriers as necessary to protect entrants from external hazards; and

1910.146(d)(3)(vi)

Verifying that conditions in the permit space are acceptable for entry throughout the duration of an authorized entry.

1910.146(d)(4)

Provide the following equipment (specified in paragraphs (d)(4)(i) through (d)(4)(ix) of this section) at no cost to employees, maintain that equipment properly, and ensure that employees use that equipment properly:

1910.146(d)(4)(i)

Testing and monitoring equipment needed to comply with paragraph (d)(5) of this section;

1910.146(d)(4)(ii)

Ventilating equipment needed to obtain acceptable entry conditions;

1910.146(d)(4)(iii)

Communications equipment necessary for compliance with paragraphs (h)(3) and (i)(5) of this section;

1910.146(d)(4)(iv)

Personal protective equipment insofar as feasible engineering and work practice controls do not adequately protect employees;

1910.146(d)(4)(v)

Lighting equipment needed to enable employees to see well enough to work safely and to exit the space quickly in an emergency;

1910.146(d)(4)(vi)

Barriers and shields as required by paragraph (d)(3)(v) of this section.

1910.146(d)(4)(vii)

Equipment, such as ladders, needed for safe ingress and egress by authorized entrants;

1910.146(d)(4)(viii)

Rescue and emergency equipment needed to comply with paragraph (d)(9) of this section, except to the extent that the equipment is provided by rescue services; and

1910.146(d)(4)(ix)

Any other equipment necessary for safe entry into and rescue from permit spaces.

1910.146(d)(5)

Evaluate permit space conditions as follows when entry operations are conducted:

1910.146(d)(5)(i)

Test conditions in the permit space to determine if acceptable entry conditions exist before entry is authorized to begin, except that, if isolation of the space is infeasible because the space is large or is part of a continuous system (such as a sewer), pre-entry testing shall be performed to the extent feasible before entry is authorized and, if entry is authorized, entry conditions shall be continuously monitored in the areas where authorized entrants are working;

1910.146(d)(5)(ii)

Test or monitor the permit space as necessary to determine if acceptable entry conditions are being maintained during the course of entry operations; and

1910.146(d)(5)(iii)

When testing for atmospheric hazards, test first for oxygen, then for combustible gases and vapors, and then for toxic gases and vapors.

1910.146(d)(5)(iv)

Provide each authorized entrant or that employee's authorized representative an opportunity to observe the pre-entry and any subsequent testing or monitoring of permit spaces;

1910.146(d)(5)(v)

Reevaluate the permit space in the presence of any authorized entrant or that employee's authorized representative who requests that the employer conduct such reevaluation because the entrant or representative has reason to believe that the evaluation of that space may not have been adequate;

1910.146(d)(5)(vi)

Immediately provide each authorized entrant or that employee's authorized representative with the results of any testing conducted in accordance with paragraph (d) of this section.

Note: Atmospheric testing conducted in accordance with appendix B to 1910.146 would be considered as satisfying the requirements of this paragraph. For permit space operations in sewers, atmospheric testing conducted in accordance with appendix B, as supplemented by appendix E to 1910.146, would be considered as satisfying the requirements of this paragraph.

1910.146(d)(6)

Provide at least one attendant outside the permit space into which entry is authorized for the duration of entry operations;

Note: Attendants may be assigned to monitor more than one permit space provided the duties described in paragraph (i) of this section can be effectively performed for each permit space that is monitored. Likewise, attendants may be stationed at any location outside the permit space to be monitored as long as the duties described in paragraph (i) of this section can be effectively performed for each permit space that is monitored.

1910.146(d)(7)

If multiple spaces are to be monitored by a single attendant, include in the permit program the means and procedures to enable the attendant to respond to an emergency affecting one or more of the permit spaces being monitored without distraction from the attendant's responsibilities under paragraph (i) of this section;

1910.146(d)(8)

Designate the persons who are to have active roles (as, for example, authorized entrants, attendants, entry supervisors, or persons who test or monitor the atmosphere in a permit space) in entry operations, identify the duties of each such employee, and provide each such employee with the training required by paragraph (g) of this section;

1910.146(d)(9)

Develop and implement procedures for summoning rescue and emergency services, for rescuing entrants from permit spaces, for providing necessary emergency services to rescued employees, and for preventing unauthorized personnel from attempting a rescue;

1910.146(d)(10)

Develop and implement a system for the preparation, issuance, use, and cancellation of entry permits as required by this section;

1910.146(d)(11)

Develop and implement procedures to coordinate entry operations when employees of more than one employer are working simultaneously as authorized entrants in a permit space, so that employees of one employer do not endanger the employees of any other employer;

1910.146(d)(12)

Develop and implement procedures (such as closing off a permit space and canceling the permit) necessary for concluding the entry after entry operations have been completed;

1910.146(d)(13)

Review entry operations when the employer has reason to believe that the measures taken under the permit space program may not protect employees and revise the program to correct deficiencies found to exist before subsequent entries are authorized; and

Note: Examples of circumstances requiring the review of the permit space program are: any unauthorized entry of a permit space, the detection of a permit space hazard not covered by the permit, the detection of a condition prohibited by the permit, the occurrence of an injury or near-miss during entry, a change in the use or configuration of a permit space, and employee complaints about the effectiveness of the program.

1910.146(d)(14)

Review the permit space program, using the canceled permits retained under paragraph (e)(6) of this section within 1 year after each entry and revise the program as necessary, to ensure that employees participating in entry operations are protected from permit space hazards.

Note: Employers may perform a single annual review covering all entries performed during a 12-month period. If no entry is performed during a 12-month period, no review is necessary.

Appendix C to 1910.146 presents examples of permit space programs that are considered to comply with the requirements of paragraph (d) of this section.

1910.146(e)

Permit system.

1910.146(e)(1)

Before entry is authorized, the employer shall document the completion of measures required by paragraph (d)(3) of this section by preparing an entry permit.

Note: Appendix D to 1910.146 presents examples of permits whose elements are considered to comply with the requirements of this section.

1910.146(e)(2)

Before entry begins, the entry supervisor identified on the permit shall sign the entry permit to authorize entry.

1910.146(e)(3)

The completed permit shall be made available at the time of entry to all authorized entrants or their authorized representatives, by posting it at the entry portal or by any other equally effective means, so that the entrants can confirm that pre-entry preparations have been completed.

1910.146(e)(4)

The duration of the permit may not exceed the time required to complete the assigned task or job identified on the permit in accordance with paragraph (f)(2) of this section.

1910.146(e)(5)

The entry supervisor shall terminate entry and cancel the entry permit when:

1910.146(e)(5)(i)

The entry operations covered by the entry permit have been completed; or

1910.146(e)(5)(ii)

A condition that is not allowed under the entry permit arises in or near the permit space.

1910.146(e)(6)

The employer shall retain each canceled entry permit for at least 1 year to facilitate the review of the permit-required confined space program required by paragraph (d)(14) of this section. Any problems encountered during an entry operation shall be noted on the pertinent permit so that appropriate revisions to the permit space program can be made.

1910.146(f)

Entry permit. The entry permit that documents compliance with this section and authorizes entry to a permit space shall identify:

1910.146(f)(1)

The permit space to be entered;

1910.146(f)(2)

The purpose of the entry;

1910.146(f)(3)

The date and the authorized duration of the entry permit;

1910.146(f)(4)

The authorized entrants within the permit space, by name or by such other means (for example, through the use of rosters or tracking systems) as will enable the attendant to determine quickly and accurately, for the duration of the permit, which authorized entrants are inside the permit space;

Note: This requirement may be met by inserting a reference on the entry permit as to the means used, such as a roster or tracking system, to keep track of the authorized entrants within the permit space.

1910.146(f)(5)

The personnel, by name, currently serving as attendants;

1910.146(f)(6)

The individual, by name, currently serving as entry supervisor, with a space for the signature or initials of the entry supervisor who originally authorized entry;

1910.146(f)(7)

The hazards of the permit space to be entered;

1910.146(f)(8)

The measures used to isolate the permit space and to eliminate or control permit space hazards before entry;

Note: Those measures can include the lockout or tagging of equipment and procedures for purging, inerting, ventilating, and flushing permit spaces.

1910.146(f)(9)

The acceptable entry conditions;

1910.146(f)(10)

The results of initial and periodic tests performed under paragraph (d)(5) of this section, accompanied by the names or initials of the testers and by an indication of when the tests were performed;

1910.146(f)(11)

The rescue and emergency services that can be summoned and the means (such as the equipment to use and the numbers to call) for summoning those services;

1910.146(f)(12)

The communication procedures used by authorized entrants and attendants to maintain contact during the entry;

1910.146(f)(13)

Equipment, such as personal protective equipment, testing equipment, communications equipment, alarm systems, and rescue equipment, to be provided for compliance with this section;

1910.146(f)(14)

Any other information whose inclusion is necessary, given the circumstances of the particular confined space, in order to ensure employee safety; and

1910.146(f)(15)

Any additional permits, such as for hot work, that have been issued to authorize work in the permit space.

1910.146(g)

Training.

1910.146(g)(1)

The employer shall provide training so that all employees whose work is regulated by this section acquire the understanding, knowledge, and skills necessary for the safe performance of the duties assigned under this section.

1910.146(g)(2)

Training shall be provided to each affected employee:

1910.146(g)(2)(i)

Before the employee is first assigned duties under this section;

1910.146(g)(2)(ii)

Before there is a change in assigned duties;

1910.146(g)(2)(iii)

Whenever there is a change in permit space operations that presents a hazard about which an employee has not previously been trained;

1910.146(g)(2)(iv)

Whenever the employer has reason to believe either that there are deviations from the permit space entry procedures required by paragraph (d)(3) of this section or that there are inadequacies in the employee's knowledge or use of these procedures.

1910.146(g)(3)

The training shall establish employee proficiency in the duties required by this section and shall introduce new or revised procedures, as necessary, for compliance with this section.

1910.146(g)(4)

The employer shall certify that the training required by paragraphs (g)(1) through (g)(3) of this section has been accomplished. The certification shall contain each employee's name, the signatures or initials of the trainers, and the dates of training. The certification shall be available for inspection by employees and their authorized representatives.

1910.146(h)

Duties of authorized entrants. The employer shall ensure that all authorized entrants:

1910.146(h)(1)

Know the hazards that may be faced during entry, including information on the mode, signs or symptoms, and consequences of the exposure;

1910.146(h)(2)

Properly use equipment as required by paragraph (d)(4) of this section;

1910.146(h)(3)

Communicate with the attendant as necessary to enable the attendant to monitor entrant status and to enable the attendant to alert entrants of the need to evacuate the space as required by paragraph (i)(6) of this section;

1910.146(h)(4)

Alert the attendant whenever:

1910.146(h)(4)(i)

The entrant recognizes any warning sign or symptom of exposure to a dangerous situation, or

1910.146(h)(4)(ii)

The entrant detects a prohibited condition; and

1910.146(h)(5)

Exit from the permit space as quickly as possible whenever:

1910.146(h)(5)(i)

An order to evacuate is given by the attendant or the entry supervisor,

1910.146(h)(5)(ii)

The entrant recognizes any warning sign or symptom of exposure to a dangerous situation,

1910.146(h)(5)(iii)

The entrant detects a prohibited condition, or

1910.146(h)(5)(iv)

An evacuation alarm is activated.

1910.146(i)

Duties of attendants. The employer shall ensure that each attendant:

1910.146(i)(1)

Knows the hazards that may be faced during entry, including information on the mode, signs or symptoms, and consequences of the exposure;

1910.146(i)(2)

Is aware of possible behavioral effects of hazard exposure in authorized entrants;

1910.146(i)(3)

Continuously maintains an accurate count of authorized entrants in the permit space and ensures that the means used to identify authorized entrants under paragraph (f)(4) of this section accurately identifies who is in the permit space;

1910.146(i)(4)

Remains outside the permit space during entry operations until relieved by another attendant;

Note: When the employer's permit entry program allows attendant entry for rescue, attendants may enter a permit space to attempt a rescue if they have been trained and equipped for rescue operations as required by paragraph (k)(1) of this section and if they have been relieved as required by paragraph (i)(4) of this section.

1910.146(i)(5)

Communicates with authorized entrants as necessary to monitor entrant status and to alert entrants of the need to evacuate the space under paragraph (i)(6) of this section;

1910.146(i)(6)

Monitors activities inside and outside the space to determine if it is safe for entrants to remain in the space and orders the authorized entrants to evacuate the permit space immediately under any of the following conditions:

1910.146(i)(6)(i)

If the attendant detects a prohibited condition;

1910.146(i)(6)(ii)

If the attendant detects the behavioral effects of hazard exposure in an authorized entrant;

1910.146(i)(6)(iii)

If the attendant detects a situation outside the space that could endanger the authorized entrants; or

1910.146(i)(6)(iv)

If the attendant cannot effectively and safely perform all the duties required under paragraph (i) of this section;

1910.146(i)(7)

Summon rescue and other emergency services as soon as the attendant determines that authorized entrants may need assistance to escape from permit space hazards;

1910.146(i)(8)

Takes the following actions when unauthorized persons approach or enter a permit space while entry is underway:

1910.146(i)(8)(i)

Warn the unauthorized persons that they must stay away from the permit space;

1910.146(i)(8)(ii)

Advise the unauthorized persons that they must exit immediately if they have entered the permit space; and

1910.146(i)(8)(iii)

Inform the authorized entrants and the entry supervisor if unauthorized persons have entered the permit space;

1910.146(i)(9)

Performs non-entry rescues as specified by the employer's rescue procedure; and

1910.146(i)(10)

Performs no duties that might interfere with the attendant's primary duty to monitor and protect the authorized entrants.

1910.146(j)

Duties of entry supervisors. The employer shall ensure that each entry supervisor:

1910.146(j)(1)

Knows the hazards that may be faced during entry, including information on the mode, signs or symptoms, and consequences of the exposure;

1910.146(j)(2)

Verifies, by checking that the appropriate entries have been made on the permit, that all tests specified by the permit have been conducted and that all procedures and equipment specified by the permit are in place before endorsing the permit and allowing entry to begin;

1910.146(j)(3)

Terminates the entry and cancels the permit as required by paragraph (e)(5) of this section;

1910.146(j)(4)

Verifies that rescue services are available and that the means for summoning them are operable;

1910.146(j)(5)

Removes unauthorized individuals who enter or who attempt to enter the permit space during entry operations; and

1910.146(j)(6)

Determines, whenever responsibility for a permit space entry operation is transferred and at intervals dictated by the hazards and operations performed within the space, that entry operations remain consistent with terms of the entry permit and that acceptable entry conditions are maintained.

1910.146(k)

Rescue and emergency services.

1910.146(k)(1)

An employer who designates rescue and emergency services, pursuant to paragraph (d)(9) of this section, shall:

1910.146(k)(1)(i)

Evaluate a prospective rescuer's ability to respond to a rescue summons in a timely manner, considering the hazard(s) identified;

Note to paragraph (k)(1)(i): What will be considered timely will vary according to the specific hazards involved in each entry. For example, 1910.134, Respiratory Protection, requires that employers provide a standby person or persons capable of immediate action to rescue employee(s) wearing respiratory protection while in work areas defined as IDLH atmospheres.

1910.146(k)(1)(ii)

Evaluate a prospective rescue service's ability, in terms of proficiency with rescue-related tasks and equipment, to function appropriately while rescuing entrants from the particular permit space or types of permit spaces identified;

1910.146(k)(1)(iii)

Select a rescue team or service from those evaluated that:

1910.146(k)(1)(iii)(A)

Has the capability to reach the victim(s) within a time frame that is appropriate for the permit space hazard(s) identified;

1910.146(k)(1)(iii)(B)

Is equipped for and proficient in performing the needed rescue services;

1910.146(k)(1)(iv)

Inform each rescue team or service of the hazards they may confront when called on to perform rescue at the site; and

1910.146(k)(1)(v)

Provide the rescue team or service selected with access to all permit spaces from which rescue may be necessary so that the rescue service can develop appropriate rescue plans and practice rescue operations.

Note to paragraph (k)(1): Non-mandatory appendix F contains examples of criteria which employers can use in evaluating prospective rescuers as required by paragraph (k)(1) of this section.

1910.146(k)(2)

An employer whose employees have been designated to provide permit space rescue and emergency services shall take the following measures:

1910.146(k)(2)(i)

Provide affected employees with the personal protective equipment (PPE) needed to conduct permit space rescues safely and train affected employees so they are proficient in the use of that PPE, at no cost to those employees;

1910.146(k)(2)(ii)

Train affected employees to perform assigned rescue duties. The employer must ensure that such employees successfully complete the training required to establish proficiency as an authorized entrant, as provided by paragraphs (g) and (h) of this section;

1910.146(k)(2)(iii)

Train affected employees in basic first-aid and cardiopulmonary resuscitation (CPR). The employer shall ensure that at least one member of the rescue team or service holding a current certification in first aid and CPR is available; and

1910.146(k)(2)(iv)

Ensure that affected employees practice making permit space rescues at least once every 12 months, by means of simulated rescue operations in which they remove dummies, manikins, or actual persons from the actual permit spaces or from representative permit spaces. Representative permit spaces shall, with respect to opening size, configuration, and accessibility, simulate the types of permit spaces from which rescue is to be performed.

1910.146(k)(3)

To facilitate non-entry rescue, retrieval systems or methods shall be used whenever an authorized entrant enters a permit space, unless the retrieval equipment would increase the overall risk of entry or would not contribute to the rescue of the entrant. Retrieval systems shall meet the following requirements.

1910.146(k)(3)(i)

Each authorized entrant shall use a chest or full body harness, with a retrieval line attached at the center of the entrant's back near shoulder level, above the entrant's head, or at another point which the employer can establish presents a profile small enough for the successful removal of the entrant. Wristlets may be used in lieu of the chest or full body harness if the employer can demonstrate that the use of a chest or full body harness is infeasible or creates a greater hazard and that the use of wristlets is the safest and most effective alternative.

1910.146(k)(3)(ii)

The other end of the retrieval line shall be attached to a mechanical device or fixed point outside the permit space in such a manner that rescue can begin as soon as the rescuer becomes aware that rescue is necessary. A mechanical device shall be available to retrieve personnel from vertical type permit spaces more than 5 feet (1.52 m) deep

1910.146(k)(4)

If an injured entrant is exposed to a substance for which a Material Safety Data Sheet (MSDS) or other similar written information is required to be kept at the worksite, that MSDS or written information shall be made available to the medical facility treating the exposed entrant.

1910.146(l)

Employee participation.

1910.146(l)(1)

Employers shall consult with affected employees and their authorized representatives on the development and implementation of all aspects of the permit space program required by paragraph (c) of this section.

1910.146(l)(2)

Employers shall make available to affected employees and their authorized representatives all information required to be developed by this section.

[58 FR 4549, Jan. 14, 1993; 58 FR 34845, June 29, 1993; 59 FR 26115, May 19, 1994; 63 FR 66038, Dec. 1, 1998; 76 FR 80739, Dec. 27, 2011]

5.2 Confined Space Entry Letters of Interpretation by OSHA (Standard 29 CFR 1910.146)

The following note is posted on the OSHA Website directly above the Letters of Interpretation:

'OSHA requirements are set by statute, standards and regulations. Our interpretation letters explain these requirements and how they apply to particular circumstances, but they cannot create additional employer obligations. This letter constitutes OSHA's interpretation of the requirements discussed. Note that our enforcement guidance may be affected by changes to OSHA rules. Also, from time to time, we update our guidance in response to new information. To keep apprised of such developments, you can consult OSHA's website at https://www.osha.gov'.

I started this chapter by attempting to provide copies of the OSHA Letters of Interpretation regarding the Confined Space Entry regulations. Unfortunately, I found that numerous interpretations were issued, and there were too many to detail here. However, I decided to include the letters of interpretation that I felt would be most beneficial to field practitioners.

Please recognize that from the OSHA perspective, all these letters are important, including those omitted here, but available on the OSHA website, since they clearly illustrate the OSHA point of view on the regulatory requirements. All the OSHA Letters of Interpretation can easily be located on the OSHA website at the following web address:

https://www.osha.gov/laws-regs/standardinterpretations/standardnumber/1910/1910.146%20-%20Index/result

The following details some of the most relevant OSHA Letters of Interpretation regarding Confined Space Entry:

OSHA Interpretation Number 1
Clarification regarding how OSHA's directives and letters of interpretation affect the regulations or standards.

February 20, 2002

Tod A. Phillips, Esquire

Spain & Hastings

3900 Two Houston Center

909 Fannin Street

Houston, TX 77010

Dear Mr. Phillips:

This responds to your November 26, 2001 letter to the Occupational Safety and Health Administration (OSHA) and subsequent phone conversation with members of my staff. You seek clarification regarding how OSHA's directives and letters of interpretation affect the regulations or standards. We apologize for the delay in addressing your concerns.

The issue of how directives and interpretation letters affect OSHA's regulations and standards is a legal question, which is beyond the purview of this office to answer. What we can do is address why the Agency issues directives and interpretation letters.

OSHA issues directives to assure the agency's policies, procedures, and instructions concerning agency operations are communicated effectively and timely to its personnel and other affected parties (OSHA Directive ADM 8-0.3, Chapter 5, Definitions, "Directives System").

A letter of interpretation "provides supplementary guidance that clarifies how to apply to a specific workplace situation a policy or procedure disseminated through the Code of Federal Regulations or the OSHA directive system. [Interpretation letters] may not interpret the OSHAct, or establish or expand OSHA policy. [Interpretation letters] may answer questions posed by OSHA, employers, employees, or other parties." (ADM 8-0.3, Chapter 5, Definitions, "Letter of Interpretation").

OSHA requirements are set by statute, standards, and regulations. Interpretation letters explain these requirements and how they apply to particular circumstances, but they cannot create additional employer obligations. The letters constitute OSHA's interpretation of the current requirements; however, our enforcement guidance may be affected by changes to OSHA rules. Also, from time to time we update our guidance in response to new information. To keep apprised of such developments, please consult OSHA's website at http://www.osha.gov.

If you need additional information, please do not hesitate to contact us by fax at: US Department of Labor, OSHA, Directorate of Construction, [Office of Construction Standards and Compliance Guidance], fax # 202-693-1689. You can also contact us by mail at the above office, Room N3468, 200 Constitution Avenue, N.W., Washington, D.C. 20210, although there will be a delay in our receiving correspondence by mail.

Sincerely,

Russell B. Swanson, Director

Directorate of Construction

OSHA Interpretation Number 2
Determining if certain spaces are considered confined spaces by applying the Permit-Required Confined Space definition

February 8, 1996

Ms. Remi Morrissette

Vermont Yankee Nuclear Power Corporation

P.O. Box 157

Governor Hunt Road

Vernon, VT 05354

Dear Ms. Morrissette:

This is in response to your letter of June 23, requesting guidance in determining whether certain spaces would be considered confined spaces by applying the Permit-Required Confined Spaces (PRCS) standard's definition. Please accept our apology for the delay in this response.

Your letter said that the entrance to the primary, containment area is through a personnel airlock that requires closing one door before opening the inner door. You asked, "Would OSHA consider this entry way a restricted or limited means for entry or exit as defined for a confined space?"

Yes. If the airlock doors are so arranged as to prohibit both doors from being maintained in an open position at the same time, an employee within the space cannot walk out of the space without restriction. Thus, entry and egress are restricted according to the standard's definition of "confined space." (We assume, based on the information in your letter, that the primary containment chamber meets the remaining two elements of the confined space definition.)

The fact that a space is considered a confined space by 29 CFR 1910.146 does not <u>automatically</u> mean that the space is a <u>permit-required</u> confined space. Since the chamber, however, is normally inerted with nitrogen, it appears that a permit-required confined space condition exists (has potential to contain a hazardous atmosphere) even though the nitrogen is purged with air before maintenance activities.

The following comments are provided under the presumption that the containment chamber is a permit space and that the only hazard is that of oxygen deficiency.

The containment chamber must be isolated from the source of the nitrogen. Some acceptable methods are blanking or blinding the line feeding the chamber, breaking or misaligning the line leading to the chamber, a double block and bleed system installed between the nitrogen source and the chamber. If the Nuclear Regulatory Commission's requirements would not allow isolating the chamber, then an equally safe alternative will have to be used. Any such alternative would have to be approved as a variance. The local OSHA Regional Office can provide guidance on the variance procedures and applicable regulations.

Regional Administrator

U.S. DOL-OSHA

133 Portland Street, 1st Floor

Boston, MA 02114

617 565-7164

If the atmospheric hazard is eliminated and there is no potential for it to develop further, the space can be reclassified as a "non-permit confined space."

Your letter indicated that you can monitor oxygen concentrations in the chamber without having to enter and that you have an entry policy prohibiting entry if oxygen concentrations fall below 19.5 percent. Our concern is whether the non-entry monitoring can assure that all the areas within the chamber where the employees can and will be working are above 19.5 percent oxygen concentration (no pockets where the level will be significantly below 19.5 percent). If the chamber oxygen monitoring system cannot eliminate the possibility of nitrogen pocketing, then full permit entry will have to be carried out until actual worksite testing of the atmosphere proves oxygen levels to be non-hazardous.

All planned maintenance activities or maintenance activities that become necessary during the refueling shutdown must be reviewed for their potential to introduce hazards into the chamber that would create a permit-required scenario. Where a maintenance activity has a potential for introducing a hazard, policy and procedures must be developed to limit the hazardous element so a permit-required condition cannot develop.

If you have further questions on this response, please contact [the Office of General Industry Compliance Assistance at (202) 693-1850]. Again, please accept our apology for the delay.

Sincerely,

John B. Miles, Jr., Director

Directorate of Compliance Programs

OSHA Interpretation Number 3
Request to OSHA for clarification on three topics related to Confined Space Entry

1. What is a "non-permit confined space" referred to in 29 CFR 1910.146?
2. Are there OSHA regulations specific for non-permitted confined spaces? If so, I would appreciate receiving a copy.

3. When may a log be substituted for a confined space entry permit (when is a log appropriate for confined space entry)?

January 13, 1994

Alton M. McKissick

Barge, Waggoner, Sumner and Cannon

162 Third Avenue North

Nashville, Tennessee 37201

Dear Mr. McKissick:

This is in response to your letter of December 16, 1993 to James Foster asking for clarification on three matters pertaining to confined space entry.

1. As stated in 29 CFR 1910.146(b) under definitions, a non-permit confined space is a confined space that does not contain or, with respect to atmospheric hazards, have the potential to contain any hazard capable of causing death or serious physical harm.
2. All OSHA standards are candidates to serve as a basis for issuing appropriate citations for any violations which occur in non-permit confined spaces.
3. Although there is no prohibition against keeping a log of confined space entries, the standard requires a confined space entry permit. A log may never be substituted for the required permit.

We hope this will resolve your concerns about permit-required confined spaces. If you have additional questions, please contact [the Office of General Industry Compliance Assistance at (202) 693-1850].

Sincerely,

Roger A. Clark, Director

Directorate of Compliance Programs

December 16, 1993

Mr. James F. Foster

U.S. Department of Labor

Occupational Safety and Health Administration

Office of Information and Consumer Affairs

Room N3647

Washington, D.C. 20210

RE: CONFINED SPACE ENTRY

Dear Mr. Foster:

I need clarification on three questions concerning confined space entry. They are as follows:

1. What is a "non-permit confined space" referred to in 29 CFR 1910.146?
2. Are there OSHA regulations specific for non-permitted confined spaces? If so, I would appreciate receiving a copy.
3. When may a log be substituted for a confined space entry permit (when is a log appropriate for confined space entry)?

If you have any questions or need additional information, please contact me at (615) 254-1500.

Sincerely,

Alton M. McKissick, CIH

OSHA Interpretation Number 4

Request OSHA for clarifications on the Confined Space Standard relative to gas turbine operations, including classifications for entry when the turbine is offline and isolation requirements for lockout / tag out and double block and bleed valves.

August 6, 2007

Mr. Ken Wilcoxson

945 Calle Del Encanto

Las Cruces, NM 88005

Dear Mr. Wilcoxson:

Thank you for your September 19 letter to the Occupational Safety and Health Administration's (OSHA) Directorate of Enforcement Programs (DEP). Your letter has been referred to DEP's Office of General Industry Enforcement (GIE) regarding an interpretation of 29 CFR 1910.146, the Permit-required confined spaces (PRCS) standard. Your scenario, diagram, and questions have been restated below for clarity.

Scenario: A permit-required confined space consists of a combustion turbine (CT) and the exhaust duct of a GE Frame 7FA commercial combustion turbine. See diagram below. The permit space contains the following hazards: hydraulically operated diverter damper blade, rotating third-stage turbine blades, no lighting, tripping hazards from exposed studs and bolts, high pressure natural gas fuel supply, and #2 liquid fuel oil to supply the combustion turbine.

Diagram:

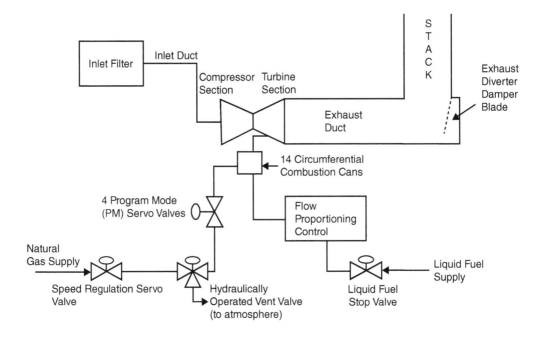

Question 1: Since the CT and exhaust duct have an actual hazardous atmosphere during normal operation, is the combustion turbine/exhaust duct/exhaust stack eligible for reclassification per 1910.146(c)(7)[1] when the CT is shut down?

Response: Reclassification under 1910.146(c)(7) is not available for permit spaces with potentially hazardous atmospheres, as appears to be the case here. (As the manufacturer's technical manual states, the valves at issue are not isolation valves and have the potential to leak.)

Question 2: When the CT is in shut-down mode, can the exhaust duct be reclassified per 1910.146(c)(7)? The National Fire Protection Association, Electric Power Research Institute, and the Combustion turbine manufacturer all indicate that there is a significant possibility of a potential hazardous atmosphere (atmosphere greater than the Lower Explosive Limit (LEL), enough to warrant that the control system and design standard mandate a purge of the CT and exhaust system.

Response: Since the permit space has a potential to contain an atmospheric hazard, reclassification per 1910.146(c)(7) would not be available.

Question 3: Would the natural gas and liquid fuel supplies be considered "flowable" sources of a potential hazardous atmosphere?

Response: The term "flowable" is found within the definition of "engulfment." The PRCS standard states, "*Engulfment* means the surrounding and effective capture of a person by a liquid or finely divided (flowable) solid substance that can be aspirated to cause death by filling or plugging the respiratory system or that can exert enough force on the body to cause death by strangulation, constriction, or crushing." Therefore, the liquid fuel supply may pose an engulfment hazard, but the natural gas would not be considered an engulfment hazard. However, natural gas supplies within a confined space may still trigger the permit space classification. Natural gas and liquid fuel can each create a hazardous atmosphere or potential hazardous atmosphere.

Question 4: When the natural gas and liquid fuel supplies are isolated[2] in accordance with 1910.146, would the space be eligible for reclassification in accordance with 1910.146(c)(7)?

Response: No, the space could not be reclassified to a non-permit space, since the space would still pose potential atmospheric hazards because, as you note, the valves are not isolation valves and may leak. In addition, the physical hazards (e.g., rotating parts, tripping hazards) in the space have not been eliminated.

Question 5: If the natural gas and liquid fuel supplies were not isolated in accordance with 1910.146, could the space be entered using the permit system?

Response: As part of the 1910.146 permit program and permit system, isolation measures must be implemented and documented. If an employer is unable to maintain acceptable entry

---

[1] 29 CFR 1910.146(c)(7)(i) states, "If the permit space poses no **actual or potential atmospheric hazards** [emphasis added] and if all hazards within the space are eliminated without entry into the space, the permit space may be reclassified as a non-permit confined space for as long as the non-atmospheric hazards remain eliminated." [back to text]

[2] 29 CFR 1910.146(b) defines isolation as "the process by which a permit space is removed from service and completely protected against the release of energy and material into the space by such means as: blanking or blinding; misaligning or removing sections of lines, pipes, or ducts; a double block and bleed system; lockout or tagout of all sources of energy; or blocking or disconnecting all mechanical linkages." [back to text]

conditions to protect the entrant from hazards or potential hazards during entry, then entry cannot be authorized.

Question 6: With regard to "isolation" and "lockout/tagout" of a flowable material, is it permissible to use solenoid valves if these devices are not capable of being locked out, i.e., no provision for attaching a lock is provided?[3]

Response: Under 1910.146, isolating a fluid flowing through a pipe by double block and bleed[4] does not specifically require that the valves be capable of being locked out. Double block and bleed is "the closure of a line, duct, or pipe by closing and *locking or tagging* [emphasis added] two in-line valves and by opening and locking or tagging a drain or vent valve in the line." Therefore, the valves and drain/vent could be tagged in accordance with the standard assuming the valves are proper for the intended application.

However, pursuant to 1910.146(d)(3), the means, procedures, and practices necessary for safe permit space entry operations must be developed and implemented. It is not clear in your scenario and diagram[5] whether the valves are opened and closed by a controller located in a different area than the valve(s). If remote operation of the valve is possible, clearly tagging the in-line valves alone would not protect an entrant from the valve being actuated remotely. Under this circumstance, additional means, procedures, or practices must be employed to assure that these valves cannot be opened (or closed as in the case of the drain) remotely.

In addition, the only appropriate isolation means that address a fluid flowing through pipes are blanking or blinding; misaligning or removing sections of lines, pipes, or ducts; or a double block and bleed system as described. Simply locking or tagging out a piping system, pursuant to 1910.147, is not appropriate for fluid isolation purposes.

As you may be aware, the State of New Mexico operates its own occupational safety and health program under a plan approved by Federal OSHA. Under this plan, the New Mexico Environment Department promulgates and enforces occupational safety and health standards under authority of State law, and posts them on its website at http://www.nmenv.state.nm.us/OHSB_website/Regulations.htm. New Mexico generally adopts standards that are identical to the Federal, and any different standards must be at least as effective as OSHA's. State Plans must interpret an identical state standard in a manner consistent with OSHA's interpretation (and/or appropriate appellate court decisions), and a different standard in a manner at least as effective as OSHA's. For information on the New Mexico permit-required confined spaces standard and its enforcement, we suggest that you contact:

Mr. Butch Tongate, Bureau Chief

Occupational Health and Safety

---

[3] An energy isolating device, such as a valve, is considered "capable of being locked out" if it meets one of the following requirements: is designed with a hasp or other part to which a lock can be attached; has a locking mechanism built into it; or can be locked without dismantling, rebuilding, or replacing the energy isolating device or permanently altering its energy control capability (such as using a lock/chain assembly on a pipeline valve or a lockable valve cover). Please note, pursuant to 1910.147, push buttons, selector switches, and other control circuit-type devices are not energy-isolating devices. [back to text]

[4] 29 CFR 1910.146(b) defines double block and bleed as "the closure of a line, duct, or pipe by closing and locking or tagging two in-line valves and by opening and locking or tagging a drain or vent valve in the line between the two closed valves." [back to text]

[5] We note that the provided diagram does not indicate that a double block and bleed system is available for the liquid fuel line. [back to text]

New Mexico Environment Department

525 Camino de los Marquez, Suite 3

Santa Fe, New Mexico 87505

Telephone: 505-876-8700

Thank you for your interest in occupational safety and health. We hope you find this information helpful. OSHA requirements are set by statute, standards, and regulations. Our interpretation letters explain these requirements and how they apply to particular circumstances, but they cannot create additional employer obligations. This letter constitutes OSHA's interpretation of the requirements discussed. Note that our enforcement guidance may be affected by changes to OSHA rules. Also, from time to time we update our guidance in response to new information. To keep apprised of such developments, you may consult OSHA's website at http://www.osha.gov If you have any further questions, please feel free to contact the OSHA Office of General Industry Enforcement at (202) 693-1850.

Sincerely,

Richard E. Fairfax, Director

Directorate of Enforcement Programs

OSHA Interpretation Number 5

OSHA Interpretation of Application of the Confined Space Standard relative to large and aging furnaces.

May 20, 2019

Mr. Anthony Green

Green Manufacturing Solutions

3000 Trailside Drive

Lexington, Kentucky 40511

Dear Mr. Green:

Thank you for your letter to the Occupational Safety and Health Administration (OSHA), regarding the definition of confined space[6] at 29 CFR 1910.146, as it may apply to large product heating and aging furnaces. The background information and questions you provided in your inquiry are paraphrased below. OSHA's responses follow your questions.

Background.

Product parts are heated in the furnaces using natural gas to temperatures ranging from 700 °F to 2200 °F to facilitate material processing. The furnace doors are actuated remotely from either the cab of a forklift or a control panel. Parts are loaded and unloaded from the furnaces using forklifts during production runs. The furnaces range in size from 8 feet 8 feet x 8 feet to 18 feet x 18 feet x 18 feet. Furnace access is provided through large garage door type units on the front of the units that move upward on a track system. The floor (hearth) of the furnaces range from 22 to 29 inches above the floor of the building depending on the size of the unit.

---

[6] In accordance with 29 CFR 1910.146, "Confined space" means a space that (1) is large enough and so configured that an employee can bodily enter and perform assigned work; (2) has limited or restricted means for entry or exit (For example, tanks, vessels, silos, storage bins, hoppers, vaults, and pits are spaces that may have limited means of entry.); and (3) is not designed for continuous employee occupancy.

Question 1: Do these types of furnaces meet the definition of a confined space? Ingress and egress only require stepping up from the building floor to the hearth of the furnace.

Response 1: The furnaces as described above do meet the definitions of a confined space, based on your letter. Stepping into furnaces that are 22 to 29 inches above the building floor, which would require more than one step (9.5 inches)[7] to climb up, may be considered restricted. Additionally, you stated that the furnaces' doors are remotely actuated. Since the furnaces' doors are actuated remotely, an employee within this space cannot walk out of the space without restriction. Thus, entry and egress are restricted according to the standard's definition of a confined space.

Question 2: Is there a threshold limit to the height of the furnace hearth above the ground that is allowed before the space would be considered "limited or restricted means for entry or exit"? For example, if the furnace is less than 30 inches above the ground, it would not be considered to have "limited or restricted means for entry or exit." Conversely, if the furnace hearth is greater than 30 inches above ground level, it would be considered "limited or restricted means for entry or exit."

Response 2: If some special means of access such as ladders, and temporary, movable, spiral, or articulated stairs are needed to enter the space, they may be considered a limited or restricted means of egress making the space confined under the standard. Therefore, as noted in our first response above, stepping into furnaces, which would require more than one step (9.5 inches) to climb up, may be considered limited or restricted means for entry or exit.

Question 3: Regardless of furnace hearth threshold height, would using a small access stool be acceptable for ingress and egress of the furnace? If used, would this then meet the definition "limited or restricted means for entry or exit"?

Response 3: Similar to the response to #2 above, the need to use an access stool would likely mean the space has a limited or restricted means for entry or exit.

Thank you for your interest in occupational safety and health. We hope you find this information helpful. OSHA requirements are set by statute, standards, and regulations. Our interpretation letters explain these requirements and how they apply to particular circumstances, but they cannot create additional employer obligations. This letter constitutes OSHA's interpretation of the requirements discussed. Note that our enforcement guidance may be affected by changes to OSHA rules. Also, from time to time, we update our guidance in response to new information. To keep apprised of such developments, you can consult OSHA's website at http://www.osha.gov. If you have any further questions, please feel free to contact the Office of General Industry and Agricultural Enforcement at (202) 693-1850.

Sincerely,

Patrick J. Kapust, Acting Director

Directorate of Enforcement Programs

---

[7] In accordance with 29 CFR 1910.25 (c)(3), have a minimum tread depth of 9.5 inches (24 cm).

OSHA Interpretation Number 6

OSHA Interpretation of the Requirement for and the Qualification of Rescue Services.

December 20, 1994

Mr. Jim Matthies, Chief

Hazelwood Fire Department

Fire Station #2

6800 Howdershell Road

Hazelwood, MO 63042

Dear Mr. Matthies:

This is in response to your letter of July 27, sent to Mr. James Foster, requesting answers to specific questions concerning the Permit-Required Confined Spaces (PRCS) standard, 29 CFR 1910.146. Your letter was assigned to the Office of General Industry Compliance Assistance for response. Please accept my apology for the delay in this response.

A copy of the Final Rule published in the *Federal Register* (January 14, 1993) and the corrections notice published on June 29, 1993, are enclosed. You may already have these publications. The actual regulatory text for the PRCS standard starts on page 4549 of the January 14, 1993 *Federal Register*.

Your seven questions/statements are answered in the order in which they were submitted. In order to be consistent with the terms used in the standard, we have substituted "permit-required confined space or permit space(s) where you have used "confined space(s)."

1. It is the employer's (facility's) responsibility to identify their permit-required confined spaces.
   Yes. The General requirements paragraph 1910.146(c)(1) requires that all employers in General Industry evaluate their workplace to determine if any permit-required confined spaces [permit space(s)] exist. In order to make a permit space determination, all spaces fitting the definition of a confined space must be identified. The standard does not require a listing of confined spaces or permit spaces; however, a prudent employer would memorialize both classes of spaces, because the standard's requirement to identify permit spaces is ongoing.
2. Companies may provide inside rescue teams only.
   No. The Rescue and emergency services paragraph 1910.146(k) allows an employer two options: (1) to have their own employees enter permit spaces to perform rescue, or (2) to arrange to have persons other than their own employees enter to perform rescue. Both of these options carry requirements for the training and education of employees who will be performing the rescue. These requirements are specified in paragraphs (k)(1) and (k)(2) of the standard.
3. In the context of the "rule" only and legal responsibilities of fire departments not withstanding, a company may ask for the fire department to become the rescue team for their confined space rescues, but the rule does not state a municipal fire department must perform that service.
   True. The PRCS standard [1910.146(k)(1)] places the obligation on the "host" employer to make arrangements for someone else (e.g., personal services contractor, fire department) to perform rescue.

4. If a fire department does provide permit space rescue service, all personnel must be cardiopulmonary resuscitation (CPR) certified and trained to advanced first aid level AND must be trained on the procedures for each specific permitted confined space AND must train in similar type spaces annually.

   The first part of your statement is partially true. Paragraph [1910.146(k)(2)(iii)] requires that each member of the rescue service (the personnel designated to rescue employees from permit spaces) must be trained in <u>basic</u> (not advanced) first aid and in CPR. **The standard requires that only one member of the rescue service present at the rescue scene need to hold a current certification in both basic first aid and CPR.** This requirement could also be met by two rescue service members, one holding a current certificate in first aid and the other holding a current certificate in CPR, and both being at the rescue scene.

   The second part of your statement is also partially true. Rescue service personnel may not have to be trained on the rescue procedures for each specific space at the host employer's facility. If there are several identically configured spaces, the spaces can be grouped for rescue training purposes. If the spaces have identical inherent hazards (e.g., physical, atmospheric), the training of rescuers on these hazards can also be combined.

   The third part of your statement concerning annual training is accurate. It should be noted, however, that an actual rescue operation taking place during the year can be applied to this annual requirement, as long as a post-rescue critique is documented, and there is remedial action with regard to problems encountered.

5. The company is responsible for providing the required training.

   No. The PRCS standard's requirement for providing training places the obligation on the employer of the employee who will be entering the permit space. With regard to rescue, if the host employer (site employer) chooses to have their own rescue service, then the host employer is required to provide all the training. When the host employer arranges for rescue services by personnel other than his or her employees, then the obligation for providing the rescue-related training is placed on the employer of the arranged-for rescue service.

   Since the host employers possess exclusive knowledge and understanding of their permit space(s), such as inherent and unique hazards of the spaces and work practices associated with the permit space, they are obligated under 1910.146(c)(8) to provide this exclusive knowledge to all contractors entering the permit space, including the arranged-for rescue service. Thus the requirement for <u>authorized entrant</u> training for rescue services responders specified in 1910.146(k)(1)(ii) must be provided by the host employer.

   The obvious next question would be: What is the minimal "authorized entrant" training needed for rescue service employees? When the training requirements of an entrant are compared to those of a rescuer, other than the attendant employing non-entry retrieval, the noticeable difference is the exclusive information on the hazards of the permit spaces to be entered, which is reflected in paragraph (h)(1), and any equipment in the permit space which could present a hazard or will be used by the rescuers, as referenced by (h)(2).

6. The company is responsible for providing required rescue equipment.

   All host employers are required to provide rescue-related equipment with respect to non-entry retrieval systems required by 1910.146 (k)(3). (A minimal retrieval system is a full body harness with a connected retrieval line worn by all authorized entrants and a mechanical device for vertical type permit spaces.) Also required by paragraph (k)(3) is the requirement to have a mechanical device available to retrieve personnel from vertical type permit spaces more than 5 feet deep.

When a host employer elects to have its own employees provide the rescue service, then all equipment associated with rescue and the rescuer's personal protection must also be provided by the host employer. The standard does not require the host employer to provide rescue-related equipment to employees employed by another employer, such as a service contractor, or the fire department.

7. The governmental body of the jurisdiction continues to have the option of what services it provides.

It is presumed by the phrasing of this statement that the "governmental body" referred to is the municipality. The PRCS standard does not address issues of contractual arrangements between two employers. The response to number 3 above may also be applicable.

If you have further questions on this correspondence, please contact Mr. Don Kallstrom in the Office of General Industry Compliance Assistance at 202 219-8031 ext. 109.

In the future, should you require further assistance on this or other OSHA related subjects, OSHA's regional and local area offices, that are closer to you and may be more attuned to local issues, are also available to respond to your needs.

Ms. Janice Barrier

Assistant Regional Administrator for Technical Support

911 Walnut Street, Room 406

Kansas City, MO 64106

816 426-5861

Mr. Denver Holt, Area Director

St. Louis Area Office

911 Washington Avenue, Room 420

St. Louis, MO 63101

814 425-4249

Sincerely,

John B. Miles, Jr., Director

Directorate of Compliance Programs

OSHA Interpretation Number 7

OSHA Interpretation of the Requirement for the Retention of the Entry Permits into Confined Spaces (period of Retention and Where Permits are to be Retained).

October 6, 1995

Mr. James J. Goumas

Vice President, Safety & Regulatory Compliance

Rust Industrial Services, Inc.

3003 Butterfield Road

Oak Brook, Illinois 60521

Dear Mr. Goumas:

This is in response to your letter dated May 25, 1995 requesting an interpretation of the provision of the Permit Required Confined Spaces Standard (29 CFR Part 1910.146(e)(6)) which requires employers to retain cancelled entry permits for at least one year.

In your letter you state that the regulations do not specify where the permits must be maintained and you raise the question as to whether it is permissible under the regulations for an employer who is also a contractor to retain the cancelled entry permit at the facility of the host employer. You also ask that if it is not so permitted, where must the employer (contractor) retain the cancelled entry permit?

The purpose of the retention of the entry permit(s) is for their review by the employer as required under paragraph (d)(14) within one year of each entry so that the written permit space program can be revised as necessary to ensure that employees participating in entry operations are protected from permit space hazards. These permits must be retained in a location so that they are available to the employer for review.

It is not necessary to retain the cancelled permits at a particular location providing the employer can demonstrate compliance with the standard. In particular, the documentation must include any information regarding problems encountered during entry operations that was recorded to comply with paragraph(e)(6) and the review must also indicate any revision of the program that resulted from such problems.

We hope this will resolve your concerns about the permit-required confined spaces record retention. If you have additional questions, please contact [the Office of General Industry Compliance Assistance at (202) 693-1850].

Sincerely,

John B. Miles, Jr., Director

Directorate of Compliance Programs

OSHA Interpretation Number 8

Mr. Michael L. Coleman

Neotronics of North America

P.O. Box 2100

Flowery Branck, GA 30542-2100

Dear Mr. Coleman:

This is in response to your request of September 16, addressed to Occupational Safety and Health Administration's (OSHA's) Deputy Regional Administrator for Region IV requesting an interpretation of 29 CFR 1910.146 concerning the recording of atmospheric test results. Your inquiry was forwarded to my office for response.

We have repeated your questions to aid other readers with the responses.

Question:

Please define OSHA's expectation with regard to maintaining written (or stored) data relative to areas to be entered and the real time, Single Time Exposure Limits (STEL), Time Weighted Averages (TWA) values of atmospheres therein?

Answer:

The Permit-Required Confined Spaces (PRCS) standard, as a generic procedural standard for work activities in permit spaces, does not address terms such as STEL and TWA. These terms related to employee health monitoring addressed by other OSHA standards.

Note: Real time, for the purposes of this response, is that time during the testing process when the direct reading instrument is viewed for the value of the substance being tested.

Regarding data OSHA expects to be maintained from the PRCS standard's position:

1. Paragraph (f)(10) requires the results of initial and periodic tests required by paragraph (d)(5) be recorded on the entry permit and maintained for 1 year. **OSHA has made a determination regarding sampling results obtained through testing PRCS atmospheres. It is enclosed for your information.**

2. Paragraph (c)(5)(i)(C) requires that the data resulting from monitoring and inspections demonstrate that the continuous forced air ventilation is maintaining the permit space safe for entry. As a performance standard, however, there is no minimum or maximum number of data entries. The preamble (Pg. 4488) of the final rule sheds light as to the quantity of data issue. It states . . . "The data required by paragraph (c)(5)(i)(C) are essential for the employer and employees, as well as OSHA, to be able to determine whether or not the space is being maintained safe for entry with the use of ventilation alone." **Thus, from a compliance position, the quantity of data being maintained must be sufficient to convince OSHA that the powered ventilation equipment and the way the fresh air is being distributed to the immediate area where the employees are or will be working is functioning properly.**

The values to be recorded on the entry permit or recorded when the atmospheric concentration needs to be documented by the standard are "real time" concentrations.

Question:

Is a user required to document actual numeric values of all three atmospheres (Real Time, STEL, and TWA) at any time?

Answer:

No, from a 29 CFR 1910.146 prospective, the only values to be documented (recorded) are real time values.

Question:

Does the expectation include written documents on the atmosphere in real time real value sense prior to entry as suggested in appendix B to 1910.146 - Procedures for Atmospheric Testing Item (2) Verification testing?

Answer:

Yes. The standard requires that employers record initial and periodic test results on the entry permit.

Question:

If a gas detector reads only real time values (and does no calculations for averaging STEL or TWA), in what way are alarm set-points impacted?

Answer:

The PRCS standard does not require alarm set-points for testing instruments.

Regarding further assistance requested, usually the answers to questions such as these (OSHA's intent and meaning of standards) can be found either on a CD-ROM titled **OSHA Regulations, Documents. and Technical Information on CD-ROM** available through the Government Printing Office or OSHA's Internet server in Salt Lake City (http://www.osha.gov/). Attached is information on these two sources.

If you have further questions on this response, please contact Mr. Don Kallstrom in the Office of Safety Compliance Assistance at 202 219-8031 x 109. For other questions regarding this or another OSHA standard or regulation, please continue to work directly with OSHA's Regional staff.

Sincerely,

John B. Miles, Jr., Director

Directorate of Compliance Programs

OSHA Interpretation Number 9

OSHA Interpretation of OSHA's requirement relative to each rescuer that enters into a permit-required confined space to rescue an entrant are required to have their own retrieval line to facilitate non-entry rescue.

November 21, 2011

Mr. Jim Kovach

4302 Root Road

North Olmsted, Ohio 44070

Dear Mr. Kovach:

Thank you for your December 8, 2010, letter, to William J. Donovan, Region V, Assistant Regional Administrator for Enforcement Programs, which was forwarded to the Office of General Industry Enforcement for an interpretation of OSHA's Permit-Required Confined Spaces standard, 29 CFR 1910.146. In addition, you emailed a member of my staff additional questions pertaining to OSHA's Permit-Required Confined Spaces standard, which were addressed over the telephone. This constitutes OSHA's interpretation only of the requirements discussed and may not be applicable to any question not delineated within your original correspondence.

Your scenario and questions are paraphrased and our responses follow.

Scenario/Background: On December 7, 2010, I attended a meeting of the Technical Advisory Committee for the Ohio Emergency Management Agency. The committee is tasked with developing curriculum for a confined space rescue course that would be approved by the Department of Homeland Security (DHS). As a result of having the course approved by DHS, it would allow other organizational search and rescue teams throughout the United States to attend the course under a grant from DHS. During the committee meeting, a question came up concerning the use of retrieval lines for rescuers entering a confined space. I have included three pages from the draft instructor's manual describing a procedure, referred to as a "belay line," that would allow rescuers to enter a confined space to locate and extract an entrant. The "belay line" would allow multiple rescuers with a chest or full-body harness to be on the same retrieval or lifeline.

Question: Does each rescuer that enters into a permit-required confined space to rescue an entrant have to have their own retrieval line to facilitate non-entry rescue?

Reply: No. The employer who designates rescue and emergency services pursuant to 29 CFR 1910.146(d)(9) shall comply with 29 CFR 1910.146(k)(1)(ii), which states:

Evaluate a prospective rescue service's ability, in terms of proficiency with rescue-related tasks and equipment, to function appropriately while rescuing entrants from the particular permit space or types of permit spaces identified.

In addition, the employer is required to select a rescue service from those evaluated per OSHA standard 29 CFR 1910.146(k)(1)(iii)(B) that "is equipped for and proficient in performing the needed rescue services."

Further, please consider OSHA standard 29 CFR 1910.146(k)(2)(i), which states:

Provide affected employees with the personal protective equipment (PPE) needed to conduct permit space rescues safely and train affected employees so they are proficient in the use of that PPE, at no cost to those employees.

Thank you for your interest in occupational safety and health. We hope you find this information helpful. Please be aware that OSHA's enforcement guidance is subject to periodic review and clarification, amplification, or correction. Such guidance could also be affected by subsequent rulemaking. In the future, should you wish to verify that the guidance provided herein remains current, you may consult OSHA's website at www.osha.gov. If you have any further questions, please feel free to contact the Office of General Industry Enforcement at (202) 693-1850.

Sincerely,

Thomas Galassi, Director

Directorate of Enforcement Programs

OSHA Interpretation Number 10

OSHA Interpretation of the OSHA's requirement relative to when permits are required for entry into a Permit Required Confined Space.

October 18, 1995

Charles M. Bessey

Senior Research Associate

Kolene Corporation

12890 Westwood Avenue

Detroit, Michigan 48223

Dear Mr. Bessey:

This is in response to your letter of November 2, 1994 concerning when permits are required for entry into a permit required confined space. The specific question is whether a permit is required for entry into a permit-required confined space (PRCS) when the entire body will not enter the space, which has thermal and chemical hazards. The size of the opening into the PRCS is three feet by three feet and it is intended that the arm be the only part of the body that enters the PRCS. Please accept our apology for the delay in this response.

When any part of the body of an entrant breaks the plane of the opening of a PRCS large enough to allow full entry, entry is considered to have occurred and a permit is required, regardless of whether there is an intent to fully enter the space. If a part of the body were placed in an opening through which the worker could not pass into the permit-required confined space, no PRCS entry will have occurred. However, if entry by only part of the body does not expose the entrant to the possibility of injury or illness then the violation may be considered a "**de minimis**" violation. (A **de minimis** violation is one in which a standard is violated, but the violation has no direct or immediate relationship to employee safety or health. These violations are documented but no citations are issued.)

Examples of situations where entry by only part of the body into a PRCS would not expose an entrant to the possibility of injury or illness are as follows:

1. An entrant reaches through the opening of a horizontal PRCS, which is so classified only because it contains exposed live electrical parts ten feet from the opening.
2. An entrant put his head through the opening of an overhead PRCS, which is so classified only because it contains unguarded rotating parts ten feet from the opening.

Examples of situations where entry by only part of the body into a PRCS can expose an entrant to the possibility of injury or illness are as follows:

1. An entrant can possibly suffer a burn while reaching into a PRCS, which is so classified because it contains a thermal hazard.
2. An entrant can possibly fall into a below grade PRCS while standing on a vertical ladder in the opening of the space, which is so classified because it contains an oxygen deficient atmosphere.
3. An entrant can possibly become unconscious as result of his head accidentally entering a PRCS while they are reaching into a PRCS, which is so classified because it contains an oxygen deficient atmosphere.

Our interpretations of how the PRCS standard would apply in these examples is limited to the specific facts of the examples. In actual situations, all relevant factors must be considered to determine whether the PRCS standard would apply and whether a citable violation occurred.

It should be noted that an employer is obligated to take appropriate steps to protect employees reaching into a space which is not a PRCS, if such an action exposes an employee to an injury or illness.

If you have any further questions on this matter, please [contact the Office of General Industry Compliance Assistance at (202) 693-1850].

Sincerely,

John B. Miles, Jr., Director

Directorate of Compliance Programs

OSHA Interpretation Number 11

OSHA Interpretation of the OSHA's requirement relative to the definition of 'Body' when describing Entry into a Permit Required Confined Space.

October 20, 1999

Michael Johnson

Oxy Vinyls, LP
Louisville Plant
Bells Lane, P.O. Box 34370
Louisville, KY 40232-4370

Dear Mr. Johnson:

Thank you for your September 28, 1999 letter to the Occupational Safety and Health Administration's (OSHA's) Office of General Industry Compliance Assistance (GICA). You have a question regarding the Permit-Required Confined Spaces standard, 29 CFR 1910.146. Your question is restated below for clarity.

Question. In terms of permit-required confined space entry, does "body" include all extremities (hands, feet, arms, and legs) or does it indicate just the head and torso?

Reply. The term "body" refers to any part of the anatomy including all extremities.

Thank you for your interest in occupational safety and health. We hope you find this information helpful. Please be aware that OSHA's enforcement guidance is subject to periodic review and clarification, amplification, or correction. Such guidance could also be affected by subsequent rulemaking. In the future, should you wish to verify that the guidance provided herein remains current, you may consult OSHA's website at http://www.osha.gov. If you have any further questions, please feel free to contact the Office of General Industry Compliance Assistance at (202) 693-1850.

Sincerely,

Richard E. Fairfax, Director

Directorate of Compliance Programs

OSHA Interpretation Number 12

OSHA Interpretation of the OSHA's requirement relative to who is required to wear the body harnesses and retrieval lines when entering a Permit Required Confined Space.

October 15, 1993

Mr. James H. Johnson

New Dimensions in Training

1293-B North 18th Street

Springfield, Oregon 97477

Dear Mr. Johnson:

Thank you for your letter of April 27, in which you requested a written interpretation as to who is required to wear the body harnesses and retrieval lines referenced in paragraph 29 CFR 1910.146(k)(3).

Paragraph 29 CFR 1910.146(k)(3) focuses on **non-entry rescue** of authorized entrants. To facilitate non-entry rescue, a retrieval system must be in place. The paragraph specifically requires except as explained below, that "retrieval systems or methods shall be used whenever an authorized entrant enters a permit space." Thus OSHA will expect **all authorized entrants** to wear retrieval devices until it is determined by the employer that a retrieval system presents a greater hazard to the entrant for the space to be entered.

There may be circumstances where using the retrieval equipment may pose a greater risk to the entrants than not using it, or the equipment would not contribute to rescue. In those situations, where a greater hazard exists, the employer may opt not to employ non-entry rescue procedures. Your attention is directed to the preamble of 29 CFR 1910.146 (pg. 4530-4531) where there is discussion on how the agency will evaluate an employer's determination whether or not a retrieval system would contribute to a rescue without increasing the overall risk of entry.

The State of Oregon will be provided a copy of this letter through our Directorate of Federal-State Operations.

If you have any further questions concerning this interpretation please contact [the Office of General Industry Compliance Assistance at (202) 693-1850].

Sincerely,

Roger A. Clark, Director

Directorate of Compliance Programs

OSHA Interpretation Number 13

OSHA Interpretation of OSHA's requirement relative to the use of signs to warn workers of a Permit Required Confined Space Entry Project.

July 22, 1998

Dear Mr. Black:

Thank you for your letter of May 21 requesting clarification of the Occupational Safety and Health Administration's (OSHA) 29 CFR 1910.146(c)(2) standard requirement to inform exposed employees of the presence of permit-required confined spaces (PRCS) in the workplace.

Paragraph (c)(2) was interpreted in OSHA instruction [CPL 2.100, May 5, 1995], and is provided below.

How will OSHA interpret the language in paragraph 1910.146(c)(2) requiring employers to inform employees of permit spaces by posting signs or "by any other equally effective means?"

Ordinarily, information about permit spaces is most effectively and economically communicated through the use of signs. Consequently, signs would be the principal method of warning under the standard. Alternative methods, such as additional training, may be used where they are truly effective in warning all employees who could reasonably be expected to enter the space. It is the employer's obligation to assure that an alternative method is at least as effective as a sign. In some cases, employers may have to provide training in addition to signs, to protect employees who do not speak English or who would have difficulty understanding or interpreting signs. (One method by which OSHA can gauge an employer's effectiveness is through random interviews of affected employees.)

If a space has a locked entry cover or panel, or an access door that can only be opened with special tools, the use of sign's may be unnecessary. If the employer ensures that all affected employees are informed about such spaces and know that they are not to be opened without taking proper precautions, including temporary signs, to restrict unexpected or unknowing

You specifically asked us: "Will you please clarify the conditions under which labels (as warnings) must be applied to entry points of PRCS? Please emphasize in your response, when in the process of coming into compliance with the standard, must labels/warnings be applied."

After April 15, 1993, once an employer has determined a permit-required confined space exists (a confined space with an actual or potential hazard), the employer is obligated by paragraph 1910.146(c)(2) to inform exposed employees either by posting a sign or other effective means. Since the purpose of this paragraph is to provide the exposed employees with information to protect them from the hazard, the "when" an employee must be informed is before the next entry.

Your letter made reference to Federal OSHA officials in Indianapolis. The State of Indiana has been ceded jurisdiction under Section 18 of the Occupational Safety and Health Act to enact and in enforce standards at least as effective as the Federal standards. You may also wish to contact them to see if the State standards for permit required confined space are indeed more stringent than the Federal PRCS standard. The address for the State of Indiana is.

Indiana Department of Labor

[Indiana Government Center - South]

402 West Washington Street, Room W195

Indianapolis, Indiana 46204

if you have further questions on this letter, please contact [the Office of General Industry Enforcement at (202) 693-1850].

Sincerely,

John B. Miles, Jr., Director

Director, Directorate of Compliance Programs

[Corrected 10/20/2006]

OSHA Interpretation Number 14

OSHA Interpretation of OSHA's requirement relative to the number of employees associated in support of a Permit Confined Space Entry and the Duties of Each Team Member in Non-IDLH and IDLH entries.

July 17, 2014

Mr. William Verhayden

Precision Industrial Maintenance, Inc.

1710 Erie Boulevard

Schenectady, New York 12308

Dear Mr. Verhayden:

Thank you for your letter to the Occupational Safety and Health Administration (OSHA) for a clarification of OSHA's Permit-Required Confined Spaces standard, 29 CFR 1910.146 (hereinafter, "the standard"). This constitutes OSHA's interpretation only of the requirements discussed and may not be applicable to any question not delineated within your original correspondence.

Your scenario and questions are paraphrased and our responses follow.

Scenario/Background #1: The definition of an "entry supervisor" at 29 CFR 1910.146(b) states:

"Entry supervisor" means the person (such as the employer, foreman, or crew chief) responsible for determining if acceptable entry conditions are present at a permit space where entry is planned, for authorizing entry and overseeing entry operations, and for terminating entry as required by this section.

NOTE: An entry supervisor also may serve as an attendant or as an authorized entrant, as long as that person is trained and equipped as required by this section for each role he or she fills. Also, the duties of entry supervisor may be passed from one individual to another during the course of an entry operation.

Question #1: Does the special role of an entry supervisor as defined in the standard at 1910.146(b) mean that the entry supervisor and attendant can be the same trained employee with the responsibilities of two roles (attendant and entry supervisor) or does the special note below the definition mean that three authorized and trained employees need to be present during the permit-required confined space (entrant, attendant, and entry supervisor)?

Reply #1: In the preamble to the final rule, OSHA discussed transferring the responsibilities of the entry supervisor from one individual to another:

The Agency anticipates that there will be many entry situations, especially if an employer has only a few employees, where the entry supervisor will serve either as the attendant or as an authorized entrant. The language of the note indicates that this is acceptable as long as the entry supervisor is trained and equipped for each role he or she fills. All pertinent requirements relating to the duties of attendants and authorized entrants would still apply to the entry supervisor who serves as an attendant or an authorized entrant. The Agency notes that the responsibilities of the entry supervisor, as revised, are set out in paragraph (j) of the final rule.

OSHA recognizes that there are circumstances, such as when the entry permit's stated duration exceeds one workshift, under which more than one person may serve as entry supervisor for a particular entry operation. The final rule does not require the employer to repeat the entry authorization process when an entry supervisor is replaced, if there is continuous direct responsibility for the entry, with direct transfer from one entry supervisor to next, and if the successor has the necessary training and performs the required duties.

58 Fed. Reg. 4473 (Jan. 14, 1993) (discussing the definition of "entry supervisor" and its note at 1910.146(b)).

The standard, at 29 CFR 1910.146(i)(10), prohibits the attendant from performing "duties that might interfere with the attendant's primary duty to monitor and protect the authorized entrants." The same individual can serve dual roles as an entry supervisor and attendant as long as the individual's duties as an entry supervisor do not interfere with his/her primary duty as an attendant.

Question #2: Can the duties of the entry supervisor be passed from one individual to another during permit entry operations?

Reply #2: The reply above explains when the standard allows the duties of the entry supervisor to be passed from one individual to another during the course of an entry operation.

Question #3: Does the standard require three individual employees to participate in permit entry operations into a non-IDLH atmosphere?

Reply #3: Not necessarily. As stated above, for a non-IDLH atmosphere permit-required confined space entry operation, an entry supervisor may have a dual role, either as attendant/supervisor or as entrant/supervisor, provided that the standard's training and equipment requirements have been met. The employer must have a plan in place to summon rescue and emergency services per

29 CFR 1910.146(d)(9). The employer must also select a rescue team or service that has the capability to reach the victim(s) within a time frame that is appropriate for the permit space hazard(s) identified per 29 CFR 1910.146(k)(1)(iii).

The rescue team or service may be onsite or offsite, employed by the employer or another employer, but they must have the capability to reach the victim(s) within the appropriate time frame, and have appropriate training and equipment per 29 CFR 1910.146(k).

If an entry rescue is necessary, the standard allows a third person to relieve the attendant so that the attendant can perform the rescue entry, provided that the employer's permit entry program allows the attendant to enter for a rescue, and if the attendant has been properly relieved and trained per 29 CFR 1910.146(i)(4) and (k)(1).

Question #4: Can a permit entry be conducted with two employees; one acting as the attendant and entry supervisor, the other acting as the entrant; assuming they are trained and equipped as required by the standard for each role for which they will be responsible?

Reply #4: In the scenario that you describe, a permit entry into a non-IDLH atmosphere can be conducted with one person performing the dual role of the attendant and entry supervisor and the second person as the authorized entrant, as long as the employer has designated a rescue team or rescue service that meets the requirements of the standard at 29 CFR 1910.146(k).

Question #5: Assume that during a non-IDLH atmosphere permit space entry operation, one worker is the entrant and the other is both the attendant and the entry supervisor. Assume that the attendant/entry supervisor is a member of the employer's onsite rescue team, and is trained annually in rescue procedures per 29 CFR 1910.146(k)(2)(iv) and has first aid and CPR training required under 29 CFR 1910.146(k)(2)(iii). Is that enough permit-required confined space training for employees to perform a PRCS entry and still maintain compliance with the entry and rescue procedures set forth in 29 CFR 1910.146?

Note: This scenario requires that the attendant operates a retrieval system that can remove the entrant from the PRCS, without entering the PRCS himself/herself, by using a retrieval system attached to the entrant. It also requires the attendant to be able to notify advance medical support such as local emergency responders prior to commencing retrieval.

Reply #5: Compliance would depend not merely on whether training is appropriate but also on whether the attendant's assignment to perform a non-entry rescue would interfere with his or her primary duty to monitor and protect the authorized entrants. The attendant has other duties as well, including the ability to summon rescue and other emergency services as soon as the attendant determines that the entrants may need assistance to escape from the permit space hazards.

Scenario/Background #2: The standard at 1910.146(k)(1)(i) states:

Evaluate a prospective rescuer's ability to respond to a rescue summons in a timely manner, considering the hazard(s) identified;

Note to paragraph (k)(l)(i): What will be considered timely will vary according to the specific hazards involved in each entry. For example, 1910.134, Respiratory Protection, requires that employers provide a standby person or persons capable of immediate action to rescue employee(s) wearing respiratory protection while in work areas defined as IDLH atmospheres.

Scenario #2: The permit-required confined space (PRCS) has a potential IDLH atmosphere. The entrant in this permit-required confined space is wearing a pressure demand or other positive-pressure self-contained breathing apparatus (SCBA), or a pressure demand or other positive pressure supplied-air respirator with auxiliary SCBA. The entrant is attached to a retrieval system (full body harness with their own individual lifeline attached to a tripod and winch) that does not require an entry rescue to retrieve the entrant from the PRCS.

Question #6: How many employees are needed and what are their duties and rescue equipment requirements?

Reply #6: OSHA standard 29 CFR 1910.134(g)(3)(iii) requires at least one employee (or more depending on the circumstances) to be located outside the IDLH atmosphere who are trained and equipped to provide effective emergency rescue. In addition, OSHA standard 29 CFR 1910.134(g)(3)(vi) requires employee(s) located outside the IDLH atmospheres to be equipped with:

1910.134(g)(3)(vi)(A)

Pressure demand or other positive pressure SCBAs, or a pressure demand or other positive pressure supplied-air respirator with auxiliary SCBA; and either

1910.134(g)(3)(vi)(B)

Appropriate retrieval equipment for removing the employee(s) who enter(s) these hazardous atmospheres where retrieval equipment would contribute to the rescue of the employee(s) and would not increase the overall risk resulting from entry; or

1910.134(g)(3)(vi)(C)

Equivalent means for rescue where retrieval equipment is not required under paragraph (g)(3)(vi)(B).

In addition to the entry supervisor and attendant, the employer would need at least one person standing by during the entry who is trained and equipped to provide an effective emergency rescue pursuant to 1910.134(g)(3)(iii).

Question #7: Is an entrant, attendant, entry supervisor, and an outside rescue service required, or can there only be an entrant, attendant, and entry supervisor with either the attendant or the entry supervisor performing the rescue operation? This question assumes that the attendant and entry supervisor are part of the employer's onsite rescue team and are trained and certified by the employer as required under 1910.146(k)(2)(iii) and (k)(2)(iv).

Reply #7: The duties of an attendant under the permit space standard allow an attendant to enter a permit space only if non-entry rescue is not possible, the attendant has been trained and equipped for rescue operations as required by paragraph (k)(1) and they have been relieved by another attendant as required by (i)(4). Separate rescue and emergency services need to be readily available during any permit-required confined space entry into an IDLH atmosphere. These individuals must be suitably equipped and capable of responding in a timely manner. Under the respirator standard, the outside personnel maintain communication with the entrant and may perform outside rescue, but are required to be trained and suitably equipped to enter the IDLH atmosphere, if needed, to provide emergency rescue.

Scenario/Background #3: The PRCS has the possibility of an IDLH atmosphere. The entrant in the PRCS with a possible IDLH atmosphere is wearing a pressure demand or other positive pressure SCBA, or a pressure demand or positive pressure supplied-air respirator with auxiliary SCBA. Further, the entrant is wearing a full body harness but is not positively hooked to a retrieval system due to entanglement issues. This scenario would require one rescuer to enter the PRCS wearing a pressure demand or other positive pressure SCBA, or a pressure demand or other positive pressure supplied-air respirator with auxiliary SCBA to aid in the rescue of the entrant.

Questions #8 and #9: How many employees are needed and what are their duties and rescue equipment requirements?

Replies #8 and #9: See the answer above. These requirements apply to IDLH atmospheres regardless of whether the rescue is performed as a non-entry rescue or an entry rescue. The non-entry rescue could suddenly turn into an entry rescue due to unforeseen hazards that would not allow a non-entry rescue to be completed.

Question #10: Can the entry supervisor, if properly trained and authorized by the employer, act as the standby onsite rescuer to enter the PRCS in the event of an entry rescue, leaving only the attendant at the PRCS entry point?

Reply #10: Please consider that OSHA has interpreted the standard to require a separate (either in-house or outside) rescue and emergency service when permit space entry operations are performed in an IDLH atmosphere. Even in permit space entry operations involving non-IDLH atmospheres, more than one rescuer may be required in permit space entry operations depending on the hazards present and the number of authorized entrants that may require rescue. The minimum number of people required to perform work that is covered by OSHA standards for permit-required confined space entry, 29 CFR 1910.146, and respiratory protection, 1910.134 will be driven by facts such as the hazards or potential hazards, the number of entrants who may require rescue and the configuration and size of the space.

Thank you for your interest in occupational safety and health. We hope you find this information helpful. Please be aware that OSHA's enforcement guidance is subject to periodic review and clarification, amplification, or correction. Such guidance could also be affected by subsequent rulemaking. In the future, should you wish to verify that the guidance provided herein remains current, you may consult OSHA's website at www.osha.gov. If you have any further questions, please feel free to contact the Office of General Industry and Agricultural Enforcement at (202) 693-1850.

Sincerely,

Thomas Galassi, Director

Directorate of Enforcement Programs

OSHA Interpretation Number 15

Clarification of OSHA's Respiratory Protection standard, 29 CFR 1910.134, specifically pertaining to paragraph 29 CFR 1910.134(i), Breathing air quality and use. May 31, 2023

Mr. Troy Robins

Suburban Manufacturing Group

10531 Dalton Avenue NE

Monticello, Minnesota 55362

Dear Mr. Robins:

Thank you for your letter to the Occupational Safety and Health Administration (OSHA), Directorate of Enforcement Programs. Your letter requested clarification of OSHA's Respiratory Protection standard, 29 CFR 1910.134, specifically pertaining to paragraph 29 CFR 1910.134(i), Breathing air quality and use. This letter constitutes OSHA's interpretation only of the requirements discussed herein and may not be applicable to any questions not delineated within your original correspondence.

Background: The Suburban Manufacturing Group manufactures filtration and drying systems for compressed air. A part of the product line is a compressed air filtration system for use with supplied air respirators (SARs). You have asked for OSHA's interpretation on addressing air quality, as well as clarification on various designs or configurations of currently marketed breathing air systems.

Please be aware that OSHA does not approve or endorse any specific equipment, products, or manufacturers' designs, therefore any questions about specific equipment or diagrams will only be answered in general terms. Below are your paraphrased questions and our responses, including feedback, when possible, on your proposed system.

Question 1: Have the specifications for Grade D air changed since its 1989 incorporation into OSHA's Respiratory Protection standard, 29 CFR 1910.134?

Response: No, OSHA's specifications for Grade D air have not changed. 29 CFR 1910.134(i)(1)(ii) specifies that Grade D air must meet the American National Standards Institute (ANSI)/Compressed Gas Association (CGA) Commodity Specification for Air, G-7.1-1989. Your letter refers to an updated consensus CGA standard, G-7.1-2018. Per OSHA's Field Operations Manual, CPL-02-00-164, if an employer chooses to comply with an amended consensus standard rather than the one referenced in a regulation – and that consensus standard provides equal to or greater protection than the standard referenced in the regulation (e.g., 29 CFR 1910.134) – the employer's deviation from the cited standard will be considered a de minimis condition. De minimis conditions are those where an employer has implemented a measure different from one specified in a standard that has no direct or immediate relationship to safety or health. See https://www.osha.gov/enforcement/directives/cpl-02-00-164/.

Question 2: What is the acceptable dryness of air?

Response: The acceptable dryness of air will vary depending on whether the breathing air supplied to respirators comes from an air cylinder versus an air compressor. When using a cylinder, employers must ensure that the moisture content in the self-contained breathing apparatus (SCBA) cylinder does not exceed a dew point of −50 °F (−45.6 °C) at one atmosphere pressure. See 29 CFR 1910.134(i)(4)(iii). However, if employers are instead using compressors to supply breathing air to respirators, the moisture content must be minimized so that the dew point at one atmosphere is 10 °F (5.56 °C) below the ambient temperature. See 29 CFR 1910.134(i)(5)(ii). ANSI/CGA G-7.1-1989, Table 2, Moisture Conversion Data, provides information on the acceptable dryness of air (converted from the dew point).

Question 3: A supplied air respirator (SAR) is being used in an ambient temperature of 70 °F and is receiving air from a compressed air system maintaining a system pressure of 120 psig. A drying system, as a part of the industrial compressed air system, is utilized to provide a dew point in the compressed air at or below 40 °F. Would this dew point meet OSHA regulations?

Response: Without more information, we cannot determine if this dew point would meet OSHA regulations. OSHA generally refrains from solving calculations for specific equipment when an OSHA representative did not collect, or was not present, when the readings were taken, because OSHA cannot ascertain that the data is complete or accurate.

In general, the moisture content of compressed air must be kept to a minimum to prevent water from freezing in valves and connections of the air supply system. Such freezing can block air lines, fittings, and pressure regulators. See 63 FR1254, January 8, 1998. To ensure low moisture content of ambient air, 29 CFR 1910.134(i)(5)(ii) requires that the dew point at one atmosphere pressure of breathing air supplied by a compressor must be 10 °F (5.56 °C) below the ambient temperature (e.g., in plant). See OSHA's Small Entity Compliance Guide for the Respiratory Protection Standard, available at https://www.osha.gov/Publications/3384small-entity-for-respiratory-protection-standard-rev.pdf, page 51. See also response to question 1.

Question 4: Does 29 CFR 1910.134(i)(4)(iii) apply to compressed air systems used for supplied air respirators?

Response: No, 29 CFR 1910.134(i)(4)(iii) only applies to the moisture content in the SCBA cylinders used to supply breathing air to respirators. By contrast, 29 CFR 1910.134(i)(5) applies to compressors.

Question 5: Does OSHA deem the requirements of 29 CFR 1910.134(i)(4)(iii) to be too dry for SARs?

Response: 29 CFR 1910.134(i)(4)(iii) governs the moisture content in SCBA cylinders while 29 CFR 1910.134(i)(5)(ii) governs the moisture content in an air compressor for airline respirators. The two systems—an air compressor that supplies air to a SAR and the supply of air in SCBA cylinders—are different, and OSHA's applicable requirements for each system must not be interchanged.

Question 6: Can "exceed" be defined, as used in 29 CFR 1910.134(i)(4)(iii)? Does it mean that the moisture content has to be at −50 °F or lower? Does it mean that the dew point cannot be below −50 °F?

Response: 29 CFR 1910.134(i)(4)(iii) states the moisture content in SCBA cylinders of breathing air cannot exceed a dew point of −50 °F (e.g., −49 °F, −48 °F, or higher temperatures). This means the dew point temperature must be at −50 °F or lower at one atmosphere, such as at −51 °F, etc.

Question 7: To achieve a −50 °F dew point at 1 atmosphere, an air system operating at 120 psig should then have a dew point of −13.4 °F at 120 psig; would this then be the minimum dew point to meet specification?

Response: As mentioned in response 3, OSHA does not generally solve calculations for specific equipment. The answer to this question would depend on whether the air system is a breathing air compressor or for an SCBA air cylinder. Please refer to the response to question 5 above.

Question 8: Can "suitable" be defined, as used in 29 CFR 1910.134(i)(5)(iii), that compressors used to supply breathing air to respirators must have "suitable in-line air-purifying sorbent beds and filters to further ensure breathing air quality. Sorbent beds and filters shall be maintained and replaced or refurbished periodically following the manufacturer's instructions?"

Response: The term "suitable" means that the compressor is capable of delivering a continuous supply of Grade D breathing air. To ensure breathing air quality, the air-purifying sorbent beds and the filters must be changed and maintained in accordance with manufacturers' instructions. Please refer to OSHA's August 3, 1998 memorandum, Questions and Answers on the Respiratory Protection Standard, for this definition and more information on suitable in-line air-purifying sorbent beds and filters.

Question 9: Would OSHA consider incorporating the specifications identified by the International Organization of Standardization (ISO) 8573-1? This ISO 8573-1 establishes clear limits regarding particle size and quantity, dew point, and oil content, and could provide a little more clarity.

Response: 29 CFR 1910.134 cannot incorporate specifications identified by ISO without proper rulemaking procedures to change the current version of the standard.

Question 10: Is there a standard for allowable materials and piping configurations? For instance: can compressors use a flow control device (such as a regulator) that restricts the flow in one direction? Is there a maximum distance allowed between the wall-mounted breathing air filtration and the operator's mask? What type of pipe or hose can be used with the breathing air filtration unit and individual respirable air connections (plumb rigid, etc.)?

Response: 29 CFR 1910.134 does not include any information or requirements specifying materials and piping configurations. As stated above, OSHA does not endorse or approve products, nor is it able to comment on the appropriateness of the various system configurations available in the market. Also, OSHA is not a testing or approval agency, and thus, does not specify the location of filters or other mechanical components of the air compressor and respirator. This is the responsibility of the National Institute for Occupational Safety and Health (NIOSH). Respirators are certified by NIOSH as a complete assembly. OSHA depends heavily upon the expertise at NIOSH to evaluate respirators and their components. NIOSH's National Personal Protective Technology Laboratory (NPPTL) provides a testing, approval, and certification program assuring respirators used in the workplace meet the standards at 42 CFR Part 84. Please note that 29 CFR 1910.134(d)(1)(ii) requires the employer to select a NIOSH-certified respirator and all parts of the respirator, including the assembly, must be configured and used in the same manner as specified in the NIOSH certification of the respirator.

If you have questions concerning respirator certification, please contact NIOSH's NPPTL Conformity Verification and Standards Development Branch at cvsdbadmin@cdc.gov or 412-386-4000. NPPTL can be contacted at 626 Cochrans Mill Road, Pittsburgh, PA 15236, 1-800-CDC-INFO (1-800-232-4636), outside the U.S. 412-386-6111, 1-888-232-6348 TTY.

Finally, note that ANSI/CGA G-7.1-1989, ANSI Z88.2, and the manufacturer, are good resources for the proper use, maintenance, and general information on respirator assembly and their components.

Question 11: At what point is the air going to a NIOSH-approved hose and respirator assembly deemed breathing air quality in order to satisfy OSHA standards?

Response: Section 29 CFR 1910.134(i) specifies OSHA's requirements for breathing air quality. For questions concerning respirator certification (i.e., air flow through the respirator assembly), please contact NIOSH's Conformity Verification and Standards Development Branch mentioned in the response to question 10 above. Compressed air sources must have the capacity to provide adequate air quality, quantity, and flow of breathing air. The NIOSH letter referenced in your letter to us, Letter to All Users of Supplied-Air Respirators Use of Unapproved Supplied-Air Respirators in the Paint Spray and Automotive Refinishing Industries, dated May 23, 1996, states, "The current regulations (42 CFR Part 84.141) require that Grade D or higher quality air be supplied to the supplied-air respirator at the point where the NIOSH approved air-supply hose connects to the respirable air source." Please note that this letter has been archived and is no longer being updated by NIOSH. OSHA recommends that you contact NIOSH's NPPTL to confirm the accuracy of the information in their archived letter.

Other good resources to use are ANSI/CGA G-7.1-1989, which has test procedures given in sections 4 and 5 of their consensus standard, and ANSI Z88.2-1992, Section 10.5.4.3, which recommends, "–[a]s part of acceptance testing, and prior to initial use, representative sampling of the compressor air output shall be performed to ensure that it complies with the requirements in sections 10.5.1 and 10.5.4. To ensure a continued high quality air supply, and to account for any distribution system contaminant input, a representative sample should be taken at distribution points."

Question 12: If the NIOSH-approved, belt-mounted, in-line filters provide the final filtration to meet the Grade D breathing air specification, without which the air would not meet the criteria, are these systems then meeting OSHA standard and NIOSH policy?

Response: OSHA is not a respirator testing and approval agency; this responsibility belongs to NIOSH NPPTL. Please contact NIOSH's NPPTL for assistance with understanding the requirements of 42 CFR Part 84.

Question 13: According to NIOSH's Respirator User Notices, only the system's NIOSH-approved hose is technically useable for breathable air as long as the hose and filter are NIOSH-approved and the required NIOSH-approved, belt-mounted, in-line carbon filter is used to remove the odors in the air system. Is this system design allowed under OSHA regulations or standards? Is there a specific OSHA regulation or standard that states an employer must have a NIOSH-approved hose to run from a wall-mounted filter to a belt filter?

Response: Please see response to question 12. OSHA also does not interpret NIOSH Respirator User Notices. 29 CFR 1910.134(d)(1)(ii) requires the employer to select a NIOSH-certified respirator. As stated in the response to question 10 above, all parts of the respirator, including the assembly, must be configured and used in the same manner as specified in the NIOSH certification of the respirator.

Question 14: What is OSHA's interpretation of a system with a standard (or NIOSH-approved) breathing (Grade D air quality) air hose connected to a belt-mounted flow control device that contains a work line, where the work line disconnect couplers are different from the couplers for the breathing air disconnect? Stated differently, if the connection points for the respirable air and pneumatic tool (work line) are separated with different couplers, would this design be considered appropriate and acceptable by OSHA? Does it allow for the potential health and safety hazard stated by NIOSH?

Response: If the respirator and the entire assembly are not NIOSH-approved, then they are not in compliance with OSHA regulations under 29 CFR 1910.134. Please also see our letters of interpretation on belt-mounted filtration units used with any NIOSH-approved supplied-air respirator to Mr. Charles E. Martin, dated September 4, 1996, and on couplings used in supplied-air respirator systems to Mr. G.T. Slay, dated April 24, 1986, that specify that NIOSH does not evaluate or approve belt mounted filtration systems that incorporate a discharge port to supply process air and breathing air couplings must be incompatible with outlets for nonrespirable worksite air or other gas systems, respectively.

Question 15: Would OSHA consider incorporating the concerns identified by NIOSH, related to system contaminants in respirable air, in its regulations?

Response: OSHA and NIOSH worked together when OSHA developed the Respiratory Protection standard. OSHA standards must go through rulemaking procedures to make changes to the current version of the standard. This issue is not on the current regulatory agenda.

Question 16: What are OSHA's requirements for the use of matched or paired hoses, masks, or hoods? Should all pieces be manufactured by the same company?

Response: Respirator parts from different manufacturers or respirator models cannot be interchanged, as this would void the NIOSH certification. See June 20, 1997, Memorandum SCBA Cylinder Interchangeability, https://www.osha.gov/laws-regs/standardinterpretations/1997-06-20-0.

As you may be aware, Minnesota is one of 29 states or territories that operate its own occupational safety and health program under a plan approved and monitored by federal OSHA. Employers in Minnesota must comply with state occupational safety and health requirements. As a condition of plan approval, State Plans are required to adopt and enforce occupational safety and health standards that are at least as effective as those promulgated by federal OSHA. A state's interpretation of their standards must also be at least as effective as federal OSHA interpretations. If you would like further information regarding Minnesota's occupational safety and health requirements, you may contact them at:

Minnesota Department of Labor and Industry

443 Lafayette Road North

St. Paul, Minnesota 55155-4307

Telephone: (651) 284-5050

https://www.osha.gov/dcsp/osp/stateprogs/minnesota.html

Thank you for your interest in occupational safety and health. I hope you find this information helpful. OSHA's requirements are set by statute, standards, and regulations. Letters of interpretation do not create new or additional requirements but rather explain these requirements and how they apply to particular circumstances. This letter constitutes OSHA's interpretation of the requirements discussed. From time to time, letters are affected when the agency updates a standard, a legal decision impacts a standard, or changes in technology affect the interpretation. To assure that you are using the correct information and guidance, please consult OSHA's website at https://www.osha.gov. If you have further questions, please feel free to contact the Office of Health Enforcement at (202) 693-2190.

Sincerely,

Kimberly A. Stille, Director

Directorate of Enforcement Programs

End of Chapter 5 Review Quiz

1 As per the US OSHA Confined Space Entry standard, attendant means an _____ stationed outside one or more permit spaces who monitors the authorized _____ and who performs all attendant's duties assigned in the employer's permit space program.

   Answer(s):

2 Permit-required confined space (permit space) means a confined space that has one or more of the following characteristics:
   A Contains or has the potential to contain a hazardous _____.
   B Contains a material that has the potential for engulfing an _____.
   C Has an internal _____ such that an entrant could be _____ or asphyxiated by inwardly converging walls or by a floor that slopes downward and tapers to a smaller cross-section.

D Contains any other recognized serious _____ or _____ hazard.

Answer(s):

3   What is the purpose of the OSHA interpretation letters?

Answer(s):

4   Can or will an OSHA interpretation letter change or create additional employer obligations?

Answer(s):

5   How does OSHA define Immediately Dangerous to Life or Health (IDLH) in the Confined Space regulation?

Answer(s):

6   In the OSHA Confined Space regulation, how do they help protect the worker from accidentally falling into an open manway or other accidental entry into a permit-required confined space or from tools or debris falling onto others who may be working in the confined space?

Answer(s):

7   How does OSHA define a non-permit confined space?

Answer(s):

8   How does US OSHA interpret the requirements for a separate and dedicated rescue service to support emergency rescue if needed during entry into a confined space? Can the entry supervisor, if properly trained and authorized by the employer, act as the standby onsite rescuer to enter the PRCS in the event of an entry rescue, leaving only the attendant at the PRCS entry point?

Answer(s):

9   How does US OSHA interpret the confined space standard regarding who is required to wear a rescue harness that is attached to a personnel retrieval system (a hoist or similar)?

Answer(s):

10   In the US OSHA confined space standard, does the term 'body' include all extremities (hands, feet, arms and legs), or does it indicate just the head and torso? What exactly does the standard mean when describing Entry into a Permit-Required Confined Space?

Answer(s):

11   What specific gas tests are required before entry can be permitted into a permit-required confined space?

Answer(s):

12   Can the employee or their authorized representative view the results of the gas tests before entry into the confined space?

Answer(s):

13 Is there a requirement for retention of the entry permit, and if so, what period of retention is required, and why is this required?

Answer(s):

14 How does US OSHA define an oxygen-deficient atmosphere and an oxygen-enriched atmosphere?

Answer(s):

Additional References

US Occupational Safety and Health Administration (OSHA), 29 CFR 1910.146 "Permit Required Confined Spaces."

US Occupational Safety and Health Administration (OSHA), OSHA Letters of Interpretation and Directives system and functions.

OSHA Letters of Interpretation and Directives system and functions. | Occupational Safety and Health Administration.

US Occupational Safety and Health Administration (OSHA), 29 CFR 1910 Standard Interpretations | Occupational Safety and Health Administration (osha.gov). https://www.osha.gov/laws-regs/standardinterpretations/standardnumber/1910

6

The Hazard of Contaminated Breathing Air and How It Can Kill

6.1 The Use of Utility Air or Instrument Air as Breathing Air Should Not Be Permitted

Another serious hazard and one that has taken lives is of a worker connecting to an incorrect supply of what was believed to be breathing air, which turned out to be a different pressurized gas. Please refer to the list of nitrogen asphyxiation incidents in the appendix and you will find a long list of these fatality incidents. This has happened at major industrial giants as well as small shops and everything in between.

For example, many fatalities have occurred by workers connecting to a nitrogen pipeline, thinking it was instrument air or plant air. This is why rules must be in place that we should never attempt to use instrument air or any other utility air supply as breathing air. Only tested and certified air from breathing air cylinders or compressed air from an oilless compressor specifically designed for breathing air should be used to ensure breathing air quality. Occupational Safety and Health Administration's (OSHA's) specifications for Grade D 29 CFR 1910.134(i)(1)(ii) specifies that Grade D air must meet the American National Standards Institute (ANSI)/Compressed Gas Association (CGA) Commodity Specification for Air, G-7.1-1989.

There is no common or conventional system for providing utility air to support operations for process needs. For example, no conventional system design is available for compressed air systems to operate pneumatic-powered equipment such as impact guns, air for instrument air systems or air to operate the pneumatic control valves to run the facility. Some facilities will have a dedicated instrument air compressor equipped with desiccant dryers or other facilities to ensure moisture is removed from the compressed air before it is sent to the instrumentation to prevent freezing of the air during winter operations. Other sites may have a large centrifugal compressor supplying air to the utility air system with a control valve bleeding off the air for their instrument air system through the dryers before going to the instrumentation.

Many of the instrument air systems, and sometimes also the utility air systems used in petroleum refineries or petrochemical plants, are backed up with a nitrogen kick-in system. In the event of a compressor failure or any other cause of low air pressure in the instrument air system, a nitrogen control valve will open to admit nitrogen into the air systems to maintain pressure in the instrument air system. In these cases, if an individual is connected to the instrument air system and using it as breathing air at that time, they can be killed by the nitrogen being admitted into the system.

Since nitrogen has no odour and there are no effects felt by the individual, the person or persons affected will not even realize that this is happening and will simply go to sleep and never wake again. This can happen if the air compressor fails, or it can happen if there is a sudden and unexpected leak or loss of containment in the instrument air system. Anything that results in low air pressure can cause the nitrogen kick-in system to open.

The reason these systems are in place is to ensure against the loss of instrument air pressure. A sudden loss of instrument air pressure can be devastating to a petroleum refinery or petrochemical plant since all instruments will go to a fail-safe position, effectively shutting down the entire process and possibly the entire facility. But again, if a worker is using the instrument or utility air system as breathing air, they will suddenly and unknowingly start receiving a high concentration of nitrogen directly into their breathing air mask. This is why I say that we should never use utility air or instrument air as breathing air at these sites, even if the site assures us that they do not use nitrogen kick-in.

- These systems are not certified as breathing air and have not been tested to ensure they meet the strict requirements for breathing air quality.
- In many cases, these systems are backed up by a nitrogen supply in the event of low air pressure. This will subject the user to nitrogen being introduced into a system being used for breathing air, which most often means certain death.

6.2 Fatalities Due to Using Blended or Manufactured Breathing Air

Other accidents have occurred by using breathing air cylinders, which were filled by breathing air suppliers using 'blended air', created by blending nitrogen and oxygen to manufacture breathing air. Incidents have occurred where the blending was off, and cylinders of primarily nitrogen were shipped to the customer, and people have died as a result. I am aware of at least two such deaths that have occurred from individuals using 'blended' air, sometimes also referred to as 'manufactured air', air that was supposed to be blended by combining nitrogen and oxygen into a breathing air cylinder that turned out to be primarily nitrogen, labelled as 'breathing air'.

My preference is to disallow the use of blended or manufactured air. Breathing air must be supplied by a compressor certified for BA, and precautions should be taken to ensure that EVERY breathing air cylinder is tested for breathing air quality. Each breathing air cylinder should be tested individually for air quality and certified to have an oxygen content of 19.5–23.5%; OSHA 1910.134(i)(1)(ii)(A) *(see note 1)*. Breathing air cylinders should be tested and tagged with a breathing air certification tag, verifying that the cylinder meets breathing air quality standards.

Each cylinder should also be tested at the receiving facility to ensure it has the correct percentage of oxygen before it is used for breathing air. Note: I also prefer a tighter oxygen content of 20.9% (+/−0.5%).

6.3 Requirements for Breathing Air to Ensure Quality (from the OHSA Technical Manual on Respiratory Protection)

- Breathing air for atmosphere-supplying respirators must be of high purity, meet quality levels for content and not exceed certain contaminant levels and moisture requirements. Compressed air, compressed oxygen, liquid air and liquid oxygen used for respiration must be in accordance with the following requirements:
    ○ Compressed and liquid oxygen must meet the United States Pharmacopoeia for medical or breathing oxygen.

○ Compressed breathing air must meet at least the requirements for Grade D breathing air as described in the ANSI/CGA Commodity Specification for Air, G-7.1-1989.

6.4 Summary of OSHA Requirements for Breathing Air Quality

The employer ensures that compressed air, compressed oxygen, liquid air and liquid oxygen meet the following specifications.

- The pharmacopoeia requirements for medical or breathing oxygen.
- The requirements for Grade D breathing air as per ANSI/CGAC Specification for Air, G-7.1-1989, including specifications outlined in Table 6.1.

6.5 Notes from the Regulation and the OSHA Technical Manual on Respiratory Protection

- Compressed oxygen must not be used in atmosphere-supplying respirators, including open circuit self-contained breathing apparatus (SCBAs), which have previously used compressed air. This prohibition is intended to prevent fires and explosions that could result if high-pressure oxygen comes into contact with oil or grease that has been introduced to the respirator or the airlines during compressed-air operations. In addition, oxygen in concentrations greater than 23.5% can only be used in equipment designed for oxygen service or distribution.
- Breathing air may be supplied to respirators from cylinders or air compressors. Where cylinders are used, they must be tested and maintained as prescribed in the Shipping Container Specification Regulations of the Department of Transportation 49 CFR (parts 173 and 178). Cylinders of purchased breathing air must have a certificate of analysis from the supplier stating that the air meets the requirements for Grade D breathing air.
- The moisture content of the compressed air in the cylinder cannot exceed a dew point of −50 °F (−45.6 °C) at 1 atmosphere pressure. This requirement will prevent respirator valves from freezing, which can occur when excess moisture accumulates on the valves. All breathing gas containers must be marked in accordance with the NIOSH respirator certification standard, 42 CFR part 84.

Table 6.1 OSHA 1910.134 breathing air quality.

Oxygen content	19.5–23.5%
Hydrocarbon (condensed) content	5 mg per cubic metre of air or less.
Carbon monoxide (CO) content	10 ppm or less.
Carbon dioxide content	1,000 ppm or less.
The lack of a noticeable odour	
Moisture content (cylinder)	Does not exceed a dew point of −50 °F (−45.6 °C).
Moisture content (compressor)	Dew point at 1 atmosphere pressure is 10 °F (5.56 °C) below the ambient temperature.

6.6 Other Specific Requirements

Where compressors are used to supply breathing air, the compressor must be constructed and situated so that contaminated air cannot enter the air supply system. The location of the air intake is very important. It must be in an uncontaminated area where exhaust gases from nearby vehicles, the internal combustion engine that is powering the compressor itself (if applicable), or other exhaust gases being ventilated from the plant will not be picked up by the compressor air intake.

1. In addition, compressors must be equipped with suitable in-line, air-purifying sorbent beds and filters to further ensure breathing air quality and to minimize moisture content so that the dew point at 1 atmosphere pressure is 10 °F (5.6 °C) below the ambient temperature. Sorbent beds and filters must be maintained and replaced or refurbished periodically according to the manufacturer's recommendations, and a tag must be kept at the compressor indicating the most recent change date and the signature of the person authorized by the employer to perform the change.

2. For compressors that are not oil-lubricated, the employer must ensure that carbon monoxide levels do not exceed 10 ppm. This requirement can be met by several different methods, including the use of continuous carbon monoxide alarms, carbon monoxide sorbent materials, proper air intake location in an area free of contaminants, frequent monitoring of air quality, or the use of high-temperature alarms and automatic shutoff devices, as appropriate. Employers have flexibility in selecting the method(s) most appropriate for conditions in their workplace. Since no single method will be appropriate in all situations, several methods may be needed. For example, it may be necessary to combine the use of a carbon monoxide alarm with a carbon monoxide sorbent bed where conditions are such that a reliable carbon monoxide-free area for air intake cannot be found.

3. Oil-lubricated compressors can produce carbon monoxide (CO) if the oil enters the combustion chamber and is ignited. This problem can be particularly severe in older compressors with worn piston rings and cylinders. Consequently, if an oil-lubricated compressor is used, it must have a high-temperature or carbon monoxide alarm, or both, to monitor carbon monoxide levels. If only a high-temperature alarm is used, the air from the compressor must be tested for carbon monoxide at intervals sufficient to prevent carbon monoxide in the breathing air from exceeding 10 ppm.

    Breathing air filters coupled with a CO monitor and alarm are readily available from safety suppliers. These are available in a single-person use case, with the air filtration and CO monitors packaged as a unit. I requested several suppliers for permission to use images of these devices in this book to no avail. However, links to several safety equipment suppliers are provided in the following Additional Resources section.

4. The employer shall ensure that breathing air couplings are incompatible with outlets for non-respirable worksite air or other gas systems. No asphyxiating substance shall be introduced into breathing air lines.

5. Breathing air couplings must be incompatible with all other site utility connections, including non-respirable plant air or other gas systems, to prevent accidental connection of air-line respirators with non-respirable gases or oxygen. This is important as it only takes one mistake, and life can be lost, and it happens suddenly and without warning. Please see Figure 6.1 for an example of a modified coupling to connect nitrogen to power pneumatic (normally air-driven) equipment.

Figure 6.1 Image of a nitrogen utility quick-connect fitting modified to connect nitrogen to air supply hoses to drive pneumatic impact guns (normally driven by compressed air).

This nitrogen utility connection was modified to operate impact guns on nitrogen instead of utility air. Some plant workers prefer to use nitrogen to drive impact guns and other pneumatically operated equipment. Nitrogen is typically drier and supplied at a higher pressure. However, this can easily introduce nitrogen into a confined space or become an asphyxiant to the equipment operator.

6. US OSHA has also advised that polyvinyl chloride pipe should not be used in the transport of compressed air in utility systems. Pressurized PVC pipe is subject to instantaneous fracture and rupture if impacted while pressurized.

Impact guns typically discharge the exhaust air directly into or very close to the operator's face, and when running the equipment on nitrogen, this can asphyxiate the operator.

End of Chapter 6 Review Quiz

1 Why is it not a good idea to connect to a petroleum refinery or petrochemical plant utility air or instrument air supply and use it as breathing air to a sandblasting hood or any other respiratory breathing air system?

   Answer(s):

2 What is the primary hazard associated with using blended air or manufactured air as breathing air?

   Answer(s):

3 In the event blended or manufactured air is used for breathing air, how should it be tested and certified before use in the field?

   Answer(s):

6 The Hazard of Contaminated Breathing Air and How It Can Kill

4 Compressed oxygen must _____ be used in atmosphere-supplying respirators, including open circuit SCBAs, which have previously used compressed air.
Answer(s):

5 Breathing air _____ must be incompatible with outlets for non-respirable plant air or other gas systems to prevent accidental servicing of air-line respirators with non-respirable gases or oxygen.
Also, no _____ substance must be allowed in the breathing air lines.
Answer(s):

6 Oil-lubricated compressors can produce _____ _____ if the oil enters the combustion chamber and is ignited.
This problem can be particularly severe in _____ compressors with worn piston rings and cylinders.
Answer(s):

7 If an oil-lubricated compressor is used to supply breathing air, it must have a _____ _____ or _____ _____ alarm, or both, to monitor carbon monoxide levels.
If only a high-temperature alarm is used, the air from the compressor must be tested for _____ _____ at intervals sufficient to prevent carbon monoxide in the breathing air from exceeding 10 ppm.
Answer(s):

8 What is an important consideration for the placement of the air compressor intake when the compressor is used to supply breathing air?
Answer(s):

9 For compressors supplying breathing air, _____ _____ and _____ must be maintained and replaced or refurbished periodically according to the manufacturer's recommendations, and a _____ must be kept at the compressor indicating the most recent change date and the signature of the person authorized by the employer to perform the change.
Answer(s):

10 What is the maximum percentage of carbon monoxide allowed when testing breathing air for quality?
Answer(s):
A 5%
B 10%
C 15%
D 20%
E 25%

11 Cylinders of purchased breathing air must have a certificate of analysis from the supplier stating that the air meets the requirements for grade D breathing air. The moisture content of the compressed air in the cylinder cannot exceed a dew point of −50 °F (−45.6 °C) at 1 atmosphere pressure.

What is the purpose of this strict moisture requirement?

Answer(s):

12 The following list includes the potential hazards and benefits of workers using high-pressure nitrogen to drive their impact guns. Please select those that are clearly hazards and could result in their death under adverse conditions.

Answer(s): Please select those that are hazardous to the workers.
A The nitrogen supply operates at higher pressure, giving the impact guns more power.
B The nitrogen is also drier, making the impact guns more effective.
C The nitrogen may asphyxiate the worker, especially if used in confined spaces or where ventilation is restricted.
D The workers can get the task done faster when using nitrogen to drive the impact guns.
E The use of nitrogen to drive impact guns violates the policy of most companies.

Additional References

US Occupational Safety and Health Administration (OSHA), List of OSHA Letters of Interpretation for 29 CFR 1910.146. https://www.osha.gov/laws-regs/standardinterpretations/standardnumber/1910/1910.146%20-%20Index/result

Industrial Hygiene Field Operations Manual Technical Manual NMCPHC-TM6290.91-2 12 MAR 2020, Compressed Breathing Air. https://www.med.navy.mil/Portals/62/Documents/NMFA/NMCPHC/root/Industrial%20Hygiene/COMPRESSED-BREATHING-AIR.pdf?ver=kIAC6P9sm-04Hl1v274pMw%3D%3D

US Occupational Safety and Health Administration (OSHA), 29 CFR 1910.134, OSHA Respiratory Protection Standard.

CGA G7.1-1989, ANSI, Compressed Gas Association. Commodity Specification for Air.

CGA G-7.1-2018 Commodity Specification for Air. Seventh Edition.

ANSI/ASSE Z88.2-2015. Practices for Respiratory Protection.

US Occupational Safety and Health Administration (OSHA), 29 CFR 1910.134, OSHA Standard Interpretations.

US Occupational Safety and Health Administration (OSHA), Questions and Answers on the Respiratory Protection Standard. https://www.osha.gov/sites/default/files/qna.pdf

National Institute for Occupational Safety and Health, Fatality Assessment and Control Evaluation (FACE), Face 9131, Laborer Dies of Carbon Monoxide Poisoning During Sandblasting Operations in Virginia.

NIOSH [1987]–NIOSH Guide to Industrial Respiratory Protection, Cincinnati, Ohio U.S. Department of Health and Human Services, Public Health Service, Centers for Disease Control, National Institute for Occupational Safety and Health. DHHS (NIOSH) Pub. No. 87–116.

Compressed Air Best Practices NIOSH and OSHA Grade D Standard Review for Supplied Air Respirators, Dan Whyman, Compressed Air Division Manager, Air and Liquid Systems, Inc. https://www.airbestpractices.com/standards/energy-management/standards-compressed-air-system-assessments/niosh-and-osha-grade-d-stand

Deaths Involving Air-line Respirators Connected to Inert Gas Sources, *American Industrial Hygiene Association Journal*. (Currently known as Journal of Occupational and Environmental Hygiene) J. B. Hudnall, A. Suruda & D. L. Campbell, Published online: 04 June 2010. https://www.tandfonline.com/doi/abs/10.1080/15298669391354289

Suppliers of Safety Equipment:

- Industrial Safety Products.
  https://www.industrialsafetyproducts.com/contact-us/
- Empire Safety and Supply.
  https://www.empiresafety.com/
- Air Systems Products.
  https://www.airsyspro.com/contact/
- Western Safety.
  sales@westernsafety.com
- Pittsburgh Spray Equipment.
  sales@pittsburghsprayequip.com

7

Most Frequent Causes of Nitrogen Asphyxiation and How to Address Them

This list is derived directly from the Appendix, which contains a long list of nitrogen asphyxiation incidents and their causes. We will step through each causal factor and discuss how they occur and what can be done to help prevent further incidents.

1. Plant workers or contractors connecting their respirators and/or sandblast hood to a facility pipeline that was in nitrogen service or to a facility pipeline in air service but at the time was backed up by nitrogen.
2. Plant workers or contractors entering a storage tank, a truck or other process vessel without verification that the atmosphere would support human life, and/or without an approved confined space entry permit, and/or without a respirator supplied with breathing-quality air.
3. An untrained worker or a worker without proper emergency response and personal protective equipment (PPE) attempting to rescue another worker who has collapsed in or near an open process vessel or a vessel being purged with nitrogen.
4. A worker working near an open process vessel that is not gas-free or a vessel being purged with nitrogen without personal protective equipment (i.e. a respirator supplied with breathing-quality air).
5. A worker working in an immediately dangerous to life or health (IDLH) environment and using a breathing air (BA) cylinder containing a gas other than BA.
6. A worker working in an IDLH environment running out of air in their self-contained breathing apparatus (SCBA).
7. A worker working in an IDLH environment with the wrong type of respirator (i.e. using an air-purifying respirator instead of breathing air).
8. Workers working in a confined space where a section of process tubing, piping or other equipment fails, resulting in the release of nitrogen.
9. Workers transporting either pressurized nitrogen or liquid nitrogen cylinders in the cab of a vehicle.

1 Plant workers or contractors connecting their respirators and/or sandblast hood to a facility pipeline that was in nitrogen service or to a facility pipeline in air service but at the time was backed up by nitrogen.

This is a tough one to solve because, at the time, the workers need to connect; generally, everything else is done, and the job is waiting on the worker. If only they would stop and give some thought to what they are about to do and recognize the seriousness of the decision they are about to make. In the industry, we think of this as a 'last-minute risk assessment'. If they were to do this, they would understand the potential consequence of connecting to an unmarked pipeline for breathing air and investigate further before making the decision to make the connection without absolute assurance that they are connecting to breathing-quality air. However, due to the track record in this critical area and the decisions that are being made in the field, we must take other actions.

Many sites have established what is known as 'cardinal safety rules', also known as 'lifesaving rules'. These are actions that can result in serious injury or the immediate loss of life or limb. Climbing into an overhead pipe rack without wearing proper fall protection is typically among the series of cardinal rules. Violating one of these rules is considered a serious offense and can result in immediate dismissal from the facility or company.

Based on the number of incidents and fatalities associated with connecting to a facility pipeline for breathing air, my belief is that this should be one of the series of cardinal safety rules. This should be communicated through routine safety briefings when personnel come onto the facility to work and should be posted throughout the site along with the other lifesaving rules. The wording should be something like, 'Due to the potential for contamination, the connection of a personal respirator to pipeline-supplied air is not allowed. Only breathing air certified as Grade D per the ANSI standards is allowed and backed up by a certificate of analysis and local quality tests'.

The Occupational Safety and Health Administration (OSHA) standard for respiratory protection 29 CFR (1910.134) requires the employer to provide a respiratory protection plan. This plan should ensure that in areas where an independent source of breathing air is required, this source is specified in the plan. The OSHA standard 29 CFR (1910.134) specifies that compressed breathing air shall meet at least the requirements for grade D breathing air described in American National Standards Institute (ANSI)/Compressed Gas Association Commodity Specification for Air, G-7.1-1989, to include oxygen content of 19.5–23.5%; hydrocarbon (condensed) content of 5 mg per cubic metre of air or less; carbon monoxide (CO) content of 10 ppm or less; carbon dioxide content of 1,000 ppm or less; and the lack of a noticeable odour.

In most cases, I would expect this to be certified and tested breathing air cylinders or direct from a certified oilless breathing air compressor.

Additionally, the utility connections should be clearly labelled and colour-coded to ensure that all workers know exactly the utility they are connecting to. Each of the four main utilities, air, steam, water and nitrogen, should be clearly labelled and colour-coded, a different colour for each utility. Each of these utilities should also be equipped with separate and unique quick connections, with different connections for each utility. This design of the quick connects should make it impossible to connect one of the utilities to a different utility. Each connection should also be equipped with a check valve or non-return valve located at the utility station to prevent backflow from the process into the utility. See Figure 7.1 for an example of typical utility connections.

Note that each of the utilities is clearly labelled and colour-coded, and each one has a unique quick-connect fitting for that utility. The nitrogen drop is about 10′ away and is not shown in this image. Due to the asphyxiation properties of nitrogen, the standard at this site located the nitrogen drop away from the other utilities. The nitrogen connection is also colour-coded to identify it as nitrogen (orange at this facility). Figure 7.2 shows a close-up image of the different and unique connections to prevent cross-connecting utilities.

2 Plant workers or contractors entering a storage tank, a truck or other process vessel without verification that the atmosphere would support human life, and/or without an approved confined space entry permit, and/or without a respirator supplied with breathing-quality air.

The requirements for entering a confined space are already well established in the OSHA federal regulation 29 CFR (1919.146). This standard spells out the requirement for establishing that the space (i) is large enough and so configured that an employee can bodily enter and perform assigned work; (ii) has limited or restricted means for entry or exit (e.g. tanks, vessels, silos, storage bins, hoppers, vaults and pits are spaces that may have limited means of entry.); and (iii) is not designed for continuous employee occupancy.

A permit-required confined space (permit space) means a confined space that has one or more of the following characteristics: (i) contains or has a potential to contain a hazardous atmosphere; (ii) contains a material that has the potential for engulfing an entrant; (iii) has an internal configuration such that an entrant could be trapped or asphyxiated by inwardly converging walls or by a floor, which slopes downward and tapers to a smaller cross-section; or (iv) contains any other recognized serious safety or health hazard. This means that most of the tanks, trucks and other process vessels are, by their nature, a permit-required confined space. The OSHA regulation requires specific gas testing of the vessel and specifies the tests to be completed. The test results must meet the OSHA standards before vessel entry is allowed.

The OSHA regulation 29 CFR (1910.146(c)(5)(ii)(C) requires "Before an employee enters the space, the internal atmosphere shall be tested, with a calibrated direct-reading instrument, for oxygen content, for flammable gases and vapors, and for potential toxic air contaminants, in that order. Any employee who enters the space, or that employee's authorized representative, shall be provided an opportunity to observe the pre-entry testing required by this paragraph."

This comes down to training to ensure that all personnel working at our facilities know the entry requirements into a confined space, which includes verifying the atmosphere and the fact that it will support human life. This includes testing for oxygen, the lower explosive limit (hydrocarbons) and specific toxic tests such as hydrogen sulphide (H_2S) before entry is allowed. A standby person is required for the full duration of entry into the vessel with duties as outlined in the regulation.

Figure 7.1 Photo of typical facility utility connections.

Figure 7.2 Close-up of the three unique quick connections. The fourth unique quick connection fitting is on the nitrogen utility connection, located about 10–15 feet from these three, and at this facility, it is painted yellow to identify it as nitrogen. This facility separates nitrogen from other utilities due to nitrogen's asphyxiation properties.

Facilities should have rigorous work permit procedures developed and rigorously enforced to ensure that each process tank or vessel is thoroughly verified as safe to enter before anyone enters the vessel. Management must ensure that there are a series of ongoing field audits with corrective actions developed for audit findings and follow-up to ensure the completion of recommendations. The OSHA regulation 29 CFR 1910.146(c)(5)(ii)(H) requires "The employer shall verify that the space is safe for entry and that the pre-entry measures required by paragraph (c)(5)(ii) of this section have been taken, through a written certification that contains the date, the location of the space, and the signature of the person providing the certification. The certification shall be made before entry and shall be made available to each employee entering the space or to that employee's authorized representative."

This is not intended to be a complete list of the requirements. For more details, please refer to the regulation.

3 An untrained worker or a worker without proper emergency response and personal protective equipment (PPE) attempting to rescue another worker who has collapsed in or near an open process vessel or a vessel being purged with nitrogen.

Unfortunately, this is another scenario that is difficult to control. It is a human urge to help someone who has entered a confined space and is in trouble. Our instinct is to jump in and provide assistance to either get them out of the vessel or to try and render aid where they are. Unfortunately, all too often, the rescuer becomes the second, or even the third, victim by becoming exposed to the same hazard as the person initially at risk or in danger.

This, too, comes down to training. We all must be trained in the proper response if we see someone or come across someone in an industrial setting where hazards such as nitrogen or hydrogen sulphide (H_2S) may be present. Nitrogen is invisible and has no odour; no human senses will detect it until the person succumbs to loss of consciousness and collapses. H_2S, on the other hand, will have a strong odour of rotten eggs until the concentration reaches about 100–150 ppm. The H_2S will deaden the olfactory nerves at this concentration, and our sense of smell completely disappears. Most people assume that the H_2S has disappeared and the area is safe. Unfortunately, at about 500–700 ppm, the person will lose consciousness and collapse and can die.

We should train our workers to take the following steps immediately if they see someone collapse or pass out in an industrial setting, such as a petroleum refinery, petrochemical plant or other such worksite where hazardous materials are present.

- If they have been trained in emergency response and are properly outfitted in emergency response protective equipment, such as a self-contained breathing apparatus (SCBA), they should respond and rescue or provide emergency assistance.
- However, if they are not trained or if they do not have emergency response PPE, they should immediately call for assistance and standby to guide the responders to the incident scene.
- Otherwise, if they are not trained, it is entirely possible that they will become a victim, which will delay the response to the person they are attempting to help who is already down.

4 A worker working near an open process vessel that is not gas-free or a vessel being purged with nitrogen without personal protective equipment (i.e. a respirator supplied with breathing-quality air).

In this case, we have a vessel that is not gas-free and may be undergoing a nitrogen purge with nitrogen venting from the top of the tank or vessel. In this case, there should be no workers near the open access plates or near the area where nitrogen or other gases are being vented.

There should be warning signs and hard barricades placed at a safe distance from the openings to prohibit personnel from getting close or attempting to peer into and perform work in the vessel. There were five such incidents listed in the Appendix, which included five workers killed as a result. In a few of those cases, the workers were overcome by the vented vapours, fell into the process vessel and died.

The OSHA regulation 29 CFR 1910.146(c)(5)(ii)(B) requires "When entrance covers are removed, the opening shall be promptly guarded by a railing, temporary cover, or other temporary barrier that will prevent an accidental fall through the opening and that will protect each employee working in the space from foreign objects entering the space."

In the Appendix, there are several cases where workers were working just above an open access plate where nitrogen was venting into the atmosphere. While in those cases, no one witnessed the workers fall into the vessel, we know there was no intent of the workers to enter the vessel, so it is likely that they were overcome and fell into the vessel. For example, at the Delaware City refinery, a worker was attempting to remove a roll of tape that was noticed in the vessel about 5 feet directly below the access plate, and nitrogen was flowing from the reactor. He was later discovered dead inside the vessel. At the Exxon Baytown refinery, a worker was working directly above a nitrogen vent on a sulphur unit. His body was later found in the bottom of the unit. At the Koch Minnesota refinery, a worker crawled under a plastic cover placed over the open top of the reactor that was under nitrogen purge. He also died.

Also, in the Appendix, there were several cases where the workers were working near the vessel opening and dropped a tool or a flashlight into the vessel. In those cases, the worker quickly, and almost without thought, jumped into the vessel and immediately collapsed and died. This is another important point to discuss with the workers. First and foremost, they shouldn't be that close to the opening while the vessel is being purged. However, they should never attempt to recover something accidentally dropped into the vessel. We must ensure that the vessel is gas-free, contains adequate oxygen and has been determined to be safe to enter. Entry requires a completely different work permit and an approved gas test approved by the appropriate authority. The deaths listed in the Appendix are completely preventable, and it is so sad to see the depth and breadth of this list.

When the work permit is issued, we should discuss it with the permit acceptor and ensure they fully understand that the area near the open man-way or where nitrogen is vented is off-limits until we have an acceptable gas test on the vessel.

If work is to be done near the open man-way or where nitrogen is venting, the workers must wear a full-face respirator with certified breathing air or a self-contained breathing apparatus.

5 A worker working in an immediately dangerous to life or health (IDLH) environment and using a breathing air (BA) cylinder containing a gas other than breathing air.

This speaks directly to how we certify the breathing air cylinders when they are received at the facility. First, however, let's talk about the company we receive our cylinders from. We should ensure that the cylinders we are using are filled with air compressed using oilless compressors certified as breathing air quality. I prefer that no 'manufactured air', air blended from nitrogen and oxygen, be allowed into our facility. I am aware of at least two cases of cylinders filled with manufactured air that have resulted in death of the workers when the cylinders were filled primarily with nitrogen. Appendix A includes one such death where the worker was killed by argon in a cylinder previously identified as breathing air.

The cylinders should be filled with compressed air and should be certified as breathing air and tested individually to ensure they are certified to be 19.5–23.5%) (as per CFR 29 1910.134(i)(1)(ii)(A). BA cylinders should be tested and tagged with a BA certification tag, verifying that the cylinder meets BA quality standards. Each cylinder should be tested again at the receiving facility to confirm air quality before it is used for breathing air. *Note: I prefer to target a little tighter oxygen content of 20.9% (±0.5%) oxygen.*

If these standards are in place and are rigorously enforced, we should not have a problem with breathing air quality.

6 A worker working in an IDLH environment running out of air in their self-contained breathing apparatus (SCBA).

The OSHA regulation 29 CFR (1910.146) requires an attendant to be positioned outside the confined space entrance and in regular contact with the worker(s) inside the space. They should periodically verify with the workers their progress and status, including discussing the remaining breathing air supply. The workers have an obligation to be constantly aware of how far they are from the exit versus the remaining breathing air (in terms of minutes to the exit). The respirators are equipped with a heads-up display indicating the remaining breathing air. The respirators also have an internal alarm (audible and vibration) indicating low breathing pressure. These alarms typically sound when the breathing supply has about 25–33% remaining; however, training for firefighters specifies that the entrant should never wait for the alarm to sound before starting their exit or develop a dependence on the alarm before exiting. The same should apply to confined space entrants.

Respirators can also have a Personal Alert Safety System (PASS) alarm. The PASS alarm was created and implemented so that the SCBA user could summon or alert others for help. This is an effective device for the confined space and attendant to use as it is another way that the entrant can alert the attendant that help may be required.

Some jobs of long duration or with multiple entrants may require more than one attendant. For example, an entry into an inert atmosphere (inert entry) with specialized contractors and highly specialized equipment, including a bank of multiple breathing air cylinders, will generally require an additional attendant as a 'bottle watch' to constantly monitor the breathing air inventory in the cylinders and switching cylinders as required to maintain a reliable and continuous flow of breathing air to the entrants. This type of task also requires highly specialized procedures for all of the workers to follow.

7 A worker working in an IDLH environment with the wrong type of respirator (i.e. using an air-purifying respirator instead of breathing air).

OSHA requires employees to attend initial and annual refresher training if they are required to wear or use a respirator. The types of respirators, including air-purifying and self-contained breathing apparatus (SCBA), and the approved uses of each should be included.

Air-purifying respirator means a respirator with an air-purifying filter, cartridge or canister that removes specific air contaminants by passing ambient air through the air-purifying element. These respirators are designed to be used for organic vapours, mists and particulates. They are not designed to be used where nitrogen has replaced the breathing air.

In an atmosphere containing nitrogen, workers must be supplied breathing-quality air, typically from compressed air cylinders that have been certified as breathing air quality of 20.9% (±0.5%) oxygen.

References to the OSHA Respiratory Protection Manual and OSHA standards are included in the Additional Resource section of this book.

8 Workers working in a confined space where a section of process tubing, piping or other equipment fails, resulting in the release of nitrogen.

This type of failure is difficult to predict and even harder to prepare for. It comes down to the employees or workers being aware of the type of material or products contained in the process piping or vessels and being prepared to evacuate immediately if a leak or release is detected.

In the following examples, I believe each worker working in a confined space containing piping, process vessels or other process equipment containing an asphyxiant should wear an oxygen monitor with both visual and audible alarms to alert them in the event of a release. From my perspective, this should be a condition of employment.

For example, an analyser technician calibrating analysers in an analyser shelter should constantly be aware that they are working in a confined space and, more than likely, the equipment will contain nitrogen to supply the analysers. In a case like this, a release may be detected only by the noise associated with the release or by the alarms on the oxygen monitor. There will be no visual or odour associated with the release. However, exposure to nitrogen will result in oxygen depletion and oxygen deprivation to the worker in a matter of minutes. In this case, the worker, upon hearing the release or sensing the alarm (audible, visual or vibration), should immediately leave the shelter and ensure that others get outside as quickly as possible. This comes down to the workers' training and awareness of their surroundings.

Should an incident like this happen, the worker should immediately evacuate the confined space and ensure that no one else enters by placing barricades or signs. Then, they should report what happened to the responsible parties for follow-up.

9 Workers transporting either pressurized nitrogen or liquid nitrogen cylinders in the cab of a vehicle.

Transporting nitrogen cylinders in an enclosed vehicle exposes those in the vehicle to extreme hazards. All it takes is one cylinder valve to leak, even slightly, and the nitrogen will reduce the oxygen content of the air in the vehicle to dangerous levels, possibly asphyxiating the occupants. Again, this comes down to training, procedures and managers' expectations. Training should ensure that workers understand the hazards and risks associated with transporting nitrogen cylinders in vehicles. Procedures should explicitly specify that nitrogen or other compressed gas cylinders shall not be transported inside a vehicle and explain why. Moreover, managers should establish policies that preclude this throughout the company and the facility.

End of Chapter 7 Review Quiz

1. At the beginning of this chapter, we discussed some of the more frequent causes of fatalities due to nitrogen asphyxiation and how they can happen. How many of these more frequent causes of nitrogen asphyxiation can you name?

   Answer(s): Please list as many of these as you can.

2. What is one thing that each worker can do to help recognize the potential hazard they may face when undertaking an entry into a confined space or about to connect to what they believe to be a supply of breathing air?

   Answer(s):

3. As a minimum, what should be in place before a worker enters any type of confined space, especially one that may contain a hazard to the worker, such as the potential for lack of oxygen, the presence of hydrocarbons or a chemical or one that has restricted access and exit or other obstructions or encumbrances?

   Answer(s): Please list as many of these as you can.

4. No part of the human _____ should be placed inside the confined space until the requirements of the OSHA confined space entry standard are followed.

   Answer(s):

5   What should a worker do if they observe another person fall or enter a confined space without authorization, and if that person has obviously collapsed?

    Answer(s): Please select the most correct answer.
    A   Immediately call for help from the trained personnel rescue team and standby to aid support if needed.
    B   Attempt to hand the person a lifeline or attach a rope to the person to help extract them from the confined space.
    C   Immediately put on a SCBA and enter the vessel to remove the person to fresh air.
    D   Take action quickly and enter the vessel to rescue their downed buddy.
    Answer(s):

6   What are the key features that should be shared by all general unit utility stations that are designed to help make use of utilities safe for all personnel?

    Answer(s): Please name as many as you can.

7   How does OSHA attempt to protect a worker from being overcome by nitrogen (oxygen deprivation) or other dangerous vapours and falling into an opening in a process vessel, or from workers inside the space from being struck by falling tools or debris?

    Answer(s):

8   What is unsafe about transporting pressurized nitrogen cylinders or other compressed gas cylinders in the cab of a pickup truck or any other vehicle enclosure such as a car trunk?

    Answer(s):

9   Why is it that an air-purifying respirator will not protect a worker against oxygen deprivation in the event it is used in a nitrogen environment?

    Answer(s):

10 What is the hazard of a worker who is working near an open process vessel that is not gas-free or a vessel being purged with nitrogen without personal protective equipment (i.e. a respirator supplied with breathing-quality air.)?

    Answer(s):

Additional Resources

US Occupational Safety and Health Administration (OSHA), 29 CFR(1910) and 29 CFR (1926), Respiratory Protection January 8, 1998, Final Rule. https://www.osha.gov/laws-regs/federalregister/1998-01-08

National Institutes of Health, Division of Technical Resources, Office of Research Facilities, Engineering Requirements for N2 and LN2 Use and Storage. https://orf.od.nih.gov/TechnicalResources/Documents/Technical%20Bulletins/21TB/Engineering%20Requirements%20for%20N2%20and%20LN2%20Use%20and%20Storage-February%202021%20Technical%20Bulletin_508.pdf

8

More on Safe Utility Connections

In Figures 7.1 and 7.2, you probably noticed that the utility station was equipped with supply lines for Nitrogen, Utility Air, Utility Water and Steam. Please note that no connection is provided for potable water (drinking water). All potable water connections are typically direct from the well or a dedicated and separate water storage tank. There should be no connections to process equipment to or from this water system other than to a utility water break tank. The utility water is typically provided via a water break tank with an air gap between the water source and the water inventory in the break tank. In the case of a utility water connection to the process, the air gap in the water break tank prevents the process from backflowing into the water supply line and contaminating the potable water or the water source, such as a well or storage tank.

Also note that all four connections are colour coded, labelled and equipped with a different and non-compatible quick connection for each utility. They are also equipped with a non-return valve or check valve to prevent backflow from the process into the utility when a hose is connected to the process. Each utility also has two bleed valves, one on each side of the check valve, to verify that the check valve is holding and the process fluid is not backing into the process. This utility configuration is what we should find at all petroleum refining and petrochemical facilities to prevent cross-connecting of the utilities. As mentioned earlier, connecting a nitrogen supply line to a respirator should be physically impossible. Also, as discussed earlier, the facility safety rule should specifically prohibit the connection of pipeline-supplied air to a respirator, adding another layer of protection to the respirator user.

Each utility hose should be properly labelled, indicating the allowable service, the rated pressure and the rated temperature. The hose should also have a current inspection tag indicating that it has been pressure tested and is serviceable. The hose typically should be the same colour or at least banded with the colour used for the colour code used at the facility for that specific utility.

It is also a good idea, especially for high-pressure steam hoses, to equip the hose couplings with a coupling retention device to help prevent a flailing hose should the coupling fail in service. These devices are sometimes referred to as a 'whip check' and serve as a safety device, which, when properly installed, secures the steam hose and steam manifold between two steam hoses or between the manifold and hose. The purpose of the Whip Check is to prevent personnel injury from a flailing hose if the connection fails. The Whip Check does not prevent loss of containment of steam but may help direct the release away from personnel. See Figure 8.1 for an example of this type of device.

Now, let's look at a few examples of what we should NOT find at our facility. I show these few images to help you understand that the following examples are not acceptable. If you see this at your facility, action is required to upgrade the connections. Otherwise, the facility may experience an incident like those reported in the case studies below and in the Appendix.

Hazards of Nitrogen and Other Inert Gases: How They Can be Safely Managed, First Edition. M. Darryl Yoes.
© 2025 John Wiley & Sons, Inc. Published 2025 by John Wiley & Sons, Inc.

8 More on Safe Utility Connections

Figure 8.1 Example of a plant utility station.

Note that this utility station gives the impression of being poorly organized and maintained. The labelling is poor, with some of the utilities not labelled at all. There are no check valves to prevent the process from backflowing into the utility.

The utility where the hose is connected is not labelled, and there is no check valve to prevent backflow. This utility station certainly needs attention.

Please refer back to Figure 6.2 in Chapter 6, and you will see an example of a Nitrogen Utility Station in use at the time of the photograph. In this example, an adapter was installed at this nitrogen utility station to allow plant-supplied nitrogen to connect to a utility air hose. In this example, while not in compliance with the safety rules, the craftsmen are using nitrogen to drive their pneumatic impact guns. Some craftsmen do this since nitrogen is dry (no moisture) and is typically at a higher pressure than air.

However, this represents an extreme safety hazard and should never be allowed. The impact guns exhaust the nitrogen almost directly into the face of the craftsmen. Also, the impact gun may be used in a confined space where asphyxiation of the craftsmen is a real possibility. Most craftsmen simply do not understand the risk they are taking when they do this. This should be highlighted in the worker's task procedures and at the start of job training.

In the next several illustrations, we attempt to show you some additional hazards that you may be able to locate and have corrected at your facility. Figure 8.2 shows two different utility quick connectors located directly on the same utility line. This should not be possible since the quick connect should be dedicated to only one utility, and that dedication should be site-wide. Please also note in the image that neither quick connection is labelled, there is no vent or drain, and no check valve to prevent backflow into the utility.

Figure 8.3 a and b are examples of a quick 'cheater connection'. These are jury-rigged connectors used to facilitate connecting one type of utility hose to a different utility connection. These should never be allowed at any of our facilities, as this is an accident just waiting to happen.

The next image, Figure 8.4, shows a whip check safety device designed to prevent a failed steam or other utility hose from flailing in the event of a coupling failure. This image illustrates the

Two different connections on the same utility line.
One for nitrogen and one for air connections!

Figure 8.2 Example of a common pipeline connected to process equipment with two dissimilar utility connections. This can easily result in the wrong utility being connected. It appears that the quick connection on the right is for an air connection, and the left connection is for connecting a nitrogen hose. Each connection should be dedicated to only one service connection. Do not allow this in your facility.

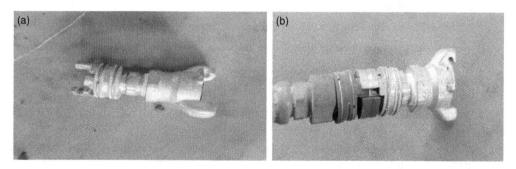

Figure 8.3 a and b Two examples of 'cheater' connections used to cross-connect two different utility hoses or connections. These should never be used; however, sometimes our technicians can be ingenious. Of course, this can result in a major incident.

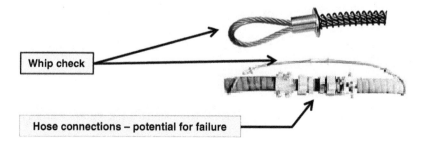

Figure 8.4 Whip check, designed to prevent a flailing hose in the event of a coupling or connection failure.

proper way to connect the whip check to each of the joined hoses and not the coupling. Please note that when properly installed, the cable loops are located as far as possible from the coupling on each side of the steam hose.

End of Chapter 8 Review Quiz

1. What is unsafe about craftsmen using nitrogen to drive impact wrenches?
   Answer(s):

2. What is unique about the typical design for potable water systems?
   Answer(s):

3. What is a common way for petroleum refining or petrochemical facilities to identify utility systems such as utility air, steam, utility water and nitrogen?
   Answer(s):

4. Is it common to see or have multiple types of quick connections installed on the same utility piping system?
   Answer(s):

5. Why is it important to have a specific quick connection for each utility that is not compatible with connections used for other utility lines?
   Answer(s): Please select the most correct answer.
   A So that an operator can easily identify the proper utility system before connecting it to process equipment.
   B To prevent cross-connecting one utility to another utility and contaminating the system.
   C To ensure that operators don't connect to the wrong utility.
   D To make it easy for the operators to keep the utility systems well organized.

6. Is it okay for operators to fabricate and use a utility cheater connection to allow connecting one utility to a different utility?
   Answer(s):

7. What is the purpose of the check valve or non-return valve, which is typically installed on utility stations?
   Answer(s):

8. What is an important aspect of utility hoses to ensure that they are properly rated and serviceable?
   Answer(s):

9. What is the purpose of a whip check device when installed on a high-pressure steam hose?
   Answer(s):

10 When the whip check is properly installed on a steam hose, should the cable loops be close to the coupling or as far from the coupling on each side as possible?

Answer(s):

11 Is it common and acceptable to have two different utility quick connections available on a common utility supply line?

Answer(s):

Additional References

Institution of Chemical Engineers (IChemE) (January 1, 2004), Hazards of Air and Oxygen. Bp Process Safety Series.

Institution of Chemical Engineers (IChemE) (January 1, 2004), Hazards of Steam. Bp Process Safety Series.

Institution of Chemical Engineers (IChemE) (January 1, 2004), Hazards of Water. Bp Process Safety Series.

Institution of Chemical Engineers (IChemE) (2009), Hazards of Nitrogen and Catalyst Handling, 2009th Edition. Bp Process Safety Series.

Institution of Chemical Engineers (IChemE) | Mar 15, 2007, Safe Handling of Light Ends 2007 Edition. Bp Process Safety Series.

This is only a few of the process safety booklets available in this series. These and others are readily available online.

9

The Hazards of Inert Entry and an Overview of the Process (Includes Case Studies of What Has Happened)

Unfortunately, we will start this discussion with a few of the reactor inert entry incidents that we are aware of occurring in our industry. These all happened while the reactors were under nitrogen purge for catalyst removal or for removal of the crust that had formed on the catalyst in the reactor. In each case, the workers were performing their inert entry tasks or working near the opening of the reactor, going about their assigned tasks when things suddenly and unexpectedly changed. In each of these tragic events, lives were changed forever.

A few examples of nitrogen-related incidents and asphyxiation incidents are highlighted below, and these are the ones we know about. All except one occurred while inert entry was ongoing.

- 1994 Conoco, Billings, MT – A Catalyst Services contract employee was removing catalyst from a reactor at the Conoco petroleum refinery. The reactor was over-pressured with nitrogen, and when he broke through the crust layer at the top of the catalyst, the pressure released forced the employee out of the reactor. He suffered multiple trauma injuries and died.
- 2001 Lavera Refinery, France – An employee was found dead at the bottom of the hydrocracker reactor under a nitrogen blanket for catalyst unloading. The employee had been making a gas test at the top manhole of the reactor, was overcome and fell inside.
- 2001 Texas City Refinery, United States – The contractor was adjusting the ladder from the top manhole of a reactor being purged with nitrogen. He was overcome and fell inside and died of asphyxiation.
- 2005 Valero Delaware City, DE – Contract fatality when a worker entered or fell into a reactor under nitrogen purge. A second fatality occurred when another worker attempted rescue (not an inert entry).
- 2011 Singapore – Inert entry contractor fatality due to breathing air cylinder contaminated with argon (one cylinder in a bank of 16 was determined to be argon).
- 2016 Placid, Port Allen, LA – Two employees were breathing air that turned out to be argon. This incident resulted in one fatality and one other employee being hospitalized with traumatic brain injury.

This is why this chapter has been included in this book: to place additional emphasis on the hazards associated with inert entry work and how they can be done safely. I don't leave you with a procedure for inert entry, but I do leave you with a list of the key activities and elements that should be included in a procedure if you are planning an inert entry at your facility for catalyst replacement or for other work or maintenance in a reactor or other vessel filled with catalyst that may be pyrophoric.

9.1 Background

Inert entry for catalyst removal has been an accepted practice in petroleum refineries and petrochemical plants for a long time. However, numerous incidents, including the death of workers, have occurred as a direct result of placing people in highly dangerous conditions, namely working in an atmosphere that is classified as Immediately Dangerous to Life and Health (IDLH). This is because the catalyst the workers are unloading or otherwise working with is activated with sulphur sites, is highly pyrophoric, and will self-ignite in the presence of air. All air is excluded from the reactor and replaced with nearly 100% nitrogen. This requires inert entry to be reinforced by a rigorously developed set of detailed procedures, and the workers are highly specialized and have specialized equipment to support workers working in this environment. Inert entry should only be used if wet dumping or catalyst passivation, which in both cases typically damages the catalyst, is not practical and if the inert entry process is deemed safe by senior management, with adequate controls in place.

Inert entry is a special type of confined space entry involving personnel entry into a confined space, typically a reactor, which contains an atmosphere that is immediately dangerous to life and health (IDLH). These tasks often involve the replacement of catalysts to help preserve the catalyst for regeneration and reuse. Most of the catalysts used in petroleum or petrochemical reactors are pyrophoric and, therefore, will self-ignite if exposed to air, even in small quantities. Also, the air in the atmosphere contains moisture, which will damage most catalysts. Thus, tasks like removing the catalyst or working with the catalyst, catalyst support systems or any reactor internals must be done in an inert atmosphere, typically nitrogen, if the catalyst is going to be regenerated and reused.

The typical requirements for confined space work must be followed. The vessel must be isolated from sources of hazardous energy and materials. Electrical sources should be isolated and locked out to prevent inadvertent re-energizing. Entry by persons not involved in the inert entry process should be eliminated. Any access to the vessel access area should be highly restricted to only those personnel engaged in the inert entry task, and they should be in full personal protective equipment (PPE), including supplied breathing air. The following should be considered when preparing for inert entry:

- Ensure adequate warning signs or inert entry work to prevent access to the work area.
- Establish barricades and personnel exclusion zones (no access to areas near the manway other than catalyst technicians in full PPE, including respiratory protection and fall prevention; harnesses or lanyards, etc.).
- Ensure adequate intrinsically safe lighting in anticipation of night work.
- Ensure that all pneumatic tools are powered by nitrogen if the tools are to be used inside the inert atmosphere (pneumatic tools should not be powered by compressed air – discharge of air into the inert space will enrich the oxygen (O_2) concentration and may result in fire).
- Establish adequate firefighting equipment at strategic locations.
- Ensure electrical bonding and grounding for equipment near vessel openings or used inside vessels.
- Ensure access to emergency rescue support (rescue personnel) and adequate provisions for rescue equipment.

Working inside a process vessel under inert atmosphere conditions involves a specialized contractor company with highly trained catalyst support staff, specially designed equipment and personnel protective equipment (PPE) for their technicians. The atmosphere inside the vessel is typically about 99% nitrogen with an oxygen concentration of about 1% (typically, 3% maximum). Most companies require a special work permit that establishes the limits for oxygen (percent), flammability (lower explosive limit [LEL], percent), hydrogen sulphide (H_2S, ppm), carbon monoxide (CO, ppm), total hydrocarbon (ppm) and benzene (ppm). The temperature in the confined space is also monitored with a maximum allowable of typically 100 °F (38 °C). Normally the testing for flammables (LEL) is conducted with the inert gas purge turned off, and testing for Oxygen is done with the inert gas flowing. Testing for flammables in an inert environment requires a specialized test device, as most LEL meters will not provide accurate readings in an inert environment.

The contractors are required to have specialized PPE, including specially designed, fully encapsulated helmets with a triple-redundant breathing supply. The helmets contain two separate cylinder-supplied hose lines delivering air supplies and an emergency escape breathing supply. The helmet is a type that is locked (or bolted) in place so the catalyst technician cannot remove the helmet while working inside the vessel. The catalyst technician inside the vessel is in radio communication with a 'hole watch' attendant in the triple air-supplied breathing apparatus and stationed at the top manway. Breathing air cylinders are typically mounted in banks of cylinders connected and monitored by a third catalyst technician, also in radio communication with the technicians (inside the vessel and the hole watch). The technician working inside the vessel will also be fitted with a special rescue harness with an attached retrieval line to aid in rescue if necessary. A separate rescue plan will be developed and ready for implementation if this becomes necessary (more on the rescue plan later).

9.2 Specialized Inert Entry Procedures

Detailed and specialized procedures must be developed covering the work inside an IDLH atmosphere and provided to the catalyst technicians, and these procedures must be enforced during the field execution. The following are only examples of the types of specialized procedures that should be in place before and during inert entry execution:

- The procedures should ensure the catalyst bed is continuously monitored for potential temperature rise. A rise of 5 °F (3 °C) in 15 minutes may indicate a potential issue with the catalyst bed or oxygen intrusion into the vessel.
- A significant amount of compressed breathing air is used during a typical inert entry procedure. One way this is done is to connect many smaller cylinders of compressed air in parallel and then use one cylinder at a time for the breathing air supplied to the catalyst technicians. The procedure should specify and enforce requirements for each cylinder, to ensure each cylinder is tested for breathing air quality before the work begins. In some past incidents, cylinders have been shipped, labelled and painted, with papers certifying them as containing breathing air, but they contained another gas or had a high percentage of nitrogen. This has resulted in on-site accidents and the deaths of catalyst technicians.
- The procedure should recognize the potential of a crust or hardened layer forming on the catalyst, which can suddenly fail due to pressure below the catalyst from the inert gas purge. This requires special precautions, such as the continuous monitoring of pressure below the catalyst bed and a temporary pressure relief valve on the inert gas line below the catalyst set at very low pressure, or maintaining the vessel in an inert state by adding nitrogen into the vessel with a hose above the catalyst bed, or similar precautions.

- Another potential issue for consideration is catalyst migration below the catalyst surface. Should the catalyst migrate, or if some catalyst has been dumped below the catalyst surface, a void can be created below the catalyst, creating a fall hazard for the catalyst technicians. The procedure should recognize these hazards and provide additional fall protection for the catalyst technicians. More on this is in the 'What Has Happened' discussion below.

The procedures must be detailed thoroughly thought out and documented before attempting to execute an inert entry. This chapter defines the minimum requirements that must be in place to protect all personnel from the hazards associated with vessels that are being purged with nitrogen or other inert gas for the purpose of inert entry. It is expected that local procedures will be more extensive and, in some cases, may even exceed the requirements provided here. A thorough review by management and local experts should be completed as an integral part of developing procedures for your facility. Procedures are typically developed based on the following elements:

9.3 Inert Entry Planning

Before the entry, the inert entry contractor must submit a detailed Site-Specific Safety Plan (e.g., a Job-Specific Safety Plan) to the facility for approval/endorsement.

The plan should include, but is not limited to, the following:

- A detailed overview of the scope of work.
- Designation of employees' roles (i.e., names of the entrant(s), attendant, air console operator, etc.) with qualifications and background, including the number of previous inert entries performed.
- A list of all specialized equipment to be used, including specialized PPE, breathing air systems, environmental monitoring equipment and rescue equipment.
- The inspection and calibration reports for the monitoring equipment to be used to support the inert entry.
- Inert gas source purity and breathing air documentation, including the requirement to provide breathing air certification documents on each breathing air cylinder before entry.
- The plans for monitoring the internal reactor temperature during all entry periods.
- A detailed communication plan, including how ongoing communication with entrants, between entrants and the attendant, and with the air console Operator, will be done.
- A site plan for barricades and warning signage, including a plan illustrating the hot zone designation. Signs should be specific to the work being done (nitrogen purging and inert confined space entry). Personnel other than those involved in the work should be excluded.
- Specify the requirements for 100% video recording and the monitoring of video camera(s) for all inert entry work.
- If required, the plan should include the lifting and rigging plans.
- The entrant rescue pre-plan (including dedicated rescue personnel and rescue equipment).

A pre-job planning meeting must be conducted before inert entry operations begin. This is to review the specific work procedures, personnel responsibilities, potential hazards and safeguards to be followed. Pre-job planning shall involve personnel responsible for the overall work on-site during the inert entry and those who will be leading the inert entry work.

9.4 The Job Safety Analysis

A thorough job safety analysis (JSA) should be carried out on each inert-entry job. The JSA should list the tasks to be done, the potential hazards associated with each task, the plan for how the hazards will be mitigated as well as who will mitigate the hazard. The JSA should include participation by all who are involved in the inert entry activities.

9.5 Acceptable Inert Atmosphere

During the entry, an acceptable oxygen concentration shall be defined as 3% oxygen or less concentration by volume. The oxygen concentration shall be continuously monitored, and alarms should be provided with audible and visual indications if the oxygen should exceed 3%. Should the oxygen exceed 3%, the workers shall be immediately evacuated from the reactor until the oxygen level is below 3%.

Pre-entry conditions must be tested and verified to be within the following parameters before the initial entry and shall be continuously monitored during all entry operations. If deviations occur within these parameters, entry operations shall cease and personnel will exit the reactor vessel until conditions return to the pre-entry status.

- The reactor's internal temperature shall be at or below 100 °F (38 °C). The reactor's internal temperature will be monitored continuously during entry operations. All entry work will cease, and entrants removed from the vessel if increases of more than 5 °F (3 °C) within 15 minutes are observed. This may be an indication of a pyrophoric reaction and a smouldering fire in the catalyst.
- Oxygen (O_2) content by volume less than 3% for initial entry and continuously thereafter.
- Combustible gases below 10% of their lower explosive limit (LEL).
  Note: special testing devices are necessary due to the inert environment (the absence of oxygen).
- Carbon monoxide (CO) is less than 50 ppm by volume (this is an indirect measurement of carbonyls, which may be toxic to the entrants).
- To help prevent worker catalyst engulfment, the catalyst height, as measured at the vessel's walls, shall not exceed 4 feet (1 meter) above the worker's feet. The workers must be trained to be alert to this hazard.
- Fall protection must be provided to the entrants in the event a void or empty space has been created in the catalyst below the workers.

The evacuation protocol must include conditions requiring personnel to evacuate the space. A mandatory evacuation is required when:

- Oxygen concentration exceeds 3% for more than 5 minutes. There shall be an immediate evacuation if the oxygen exceeds 5% at any time. This is due to the potential for a pyrophoric ignition of the catalyst.
- Evacuation in the event any of the additional pre-entry conditions for other parameters as outlined above are exceeded (see above for pre-entry conditions).

Note: API 2217A specifies additional parameters that the facility may consider. For example, other hazards may exist near the open manway outside the inert space. For example, hydrogen sulfide (H_2S), benzene or other chemicals may be present. Monitoring for these chemicals should be included in the procedure.

9.6 Inert Gas Supply and Quality

The inerting gas will typically be nitrogen, although carbon dioxide or argon may be used if approved by the facility. Whatever inert gas is used, its composition shall be a maximum of 0.5% oxygen and verified in writing before being discharged into the process vessel (the confined space). If inerting gas is supplied from a vendor's truck, the vendor shall provide the quality verification before the gas is discharged into the process vessel. All nitrogen used for vessel inerting must be verified to be 99.5% pure. Plant nitrogen may be used for the initial inerting operations, providing adequate supply and pressure to maintain an inert atmosphere.

The primary nitrogen supply during vessel entry should be a segregated system and from a reliable source. A backup nitrogen supply completely independent of the primary supply should be readily available and reliable. Entry into the vessel is only allowed when using the primary source. The plant nitrogen supply system is typically not segregated and can be contaminated due to other connections or activities. These hazards should be assessed and addressed if an independent nitrogen source is not provided.

The nitrogen primary supply pressure shall be continuously recorded during all entry operations. A means for securing nitrogen supply valves, such as locking the valves in the open position and tagging them, shall also be completed. The flow of the nitrogen supply shall also be continuously monitored during all entry operations.

9.7 Breathing Air System

The breathing air system must meet the following parameters of a full life support system, including the following guidelines. These should typically be included in the inert entry procedure and are designed to help ensure a continuous supply of certified breathing air for each entrant. The hole watch and others who may be under breathing air during the inert entry:

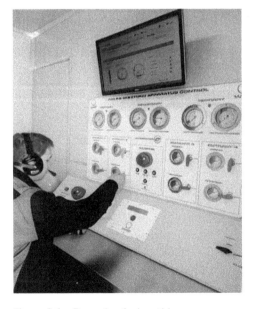

- Independent primary and secondary breathing air packs should be connected to an instrument panel (typically called the 'Breathing Air Console') to monitor the breathing air supply pressure and breathing air consumption. All breathing air isolation valves should be secured in the open position when in use and tagged with 'breathing air in use, do not close' tags. All taps used for air sampling or to vent condensate should be closed and capped when not in use. Audible alarms should be set no lower than 500 psi (35 bar) for the primary and backup air cylinders. An illustration of the breathing air console is shown in Figure 9.1 with an Inert Entry Specialist in coordination with the entrants.

- A separate air cylinder (minimum 4-hour supply) should be located at the entry manway

Figure 9.1 Example of a breathing air control panel where breathing air is continuously monitored for one or more entrants. *Source:* Courtesy of Breathe Safety Ltd. (Luno Systems).

or connected to the main control panel (depending on the type of the panel) with a pressurized hose line, typically called an 'emergency escape line' (EEL), for each entrant who is outfitted with breathing air.
- Each entrant wearing breathing air should also have an emergency escape air cylinder attached to their harness. This is a 5- to 10-minute escape cylinder and should only be used for the entrant to escape from the vessel and never to conduct work inside the vessel.
- Each entrant should be equipped with an integrated helmet/respirator assembly, referred to as a 'clamshell' or 'anti-panic shell', bolted or otherwise secured in the closed position to prevent the user from attempting removal in a panic attack scenario. Figures 9.2 and 9.3 illustrate two catalyst technicians fully outfitted with the clamshell/anti-panic shell helmets and other equipment required for entry.
- The helmet assembly should include both an independent primary air regulator and a secondary air regulator. The design should be such that if the air supply is lost to the primary regulator, the secondary regulator automatically kicks in and provides a reliable supply of breathing air to the entrant. Both regulators are of the positive pressure type to ensure against inward leakages around the face piece.
- Figure 9.2 illustrates two catalyst technicians fully suited up in the entry suits, triple redundant breathing line clamshell helmets, rescue lanyards, etc. Figure 9.3 illustrates a catalyst technician entering a reactor with assistance from the other technicians who are helping with the lifelines.
- Each entrant should also be equipped with a completely hands-free communication system that allows all entrants to communicate with each other and with the person operating the air console. Communication shall be continuous for the duration of the vessel entry. In the

Figure 9.2 Photograph of two catalyst technicians in their working suits equipped with harnesses and clamshell/anti-panic helmets with redundant breathing air and communications equipment. *Source:* Courtesy of Breathe Safety Ltd. (Luno Systems).

Figure 9.3 Photograph of two catalyst technicians in their working suits equipped with harnesses and clamshell/anti-panic helmets with redundant breathing air and communications equipment. Note: how they support each other with breathing air, unbiblical communication and rescue cables. *Source:* Courtesy of Breathe Safety Ltd. (Luno Systems).

Figure 9.4 Catalyst technician entering a permit required space with inert entry. Note the use of the clamshell helmet, communication cables and rescue harness. *Source:* Courtesy of Breathe Safety Ltd. (Luno Systems).

event of an emergency or loss of communication, there shall also be an alternative way for the entrants to communicate or provide an alert to the air monitoring console operator. Figures 9.3 and 9.4 illustrate many protective features mentioned in this section, such as the highly specialized helmets, harnesses and the 10-minute escape breathing air pack.

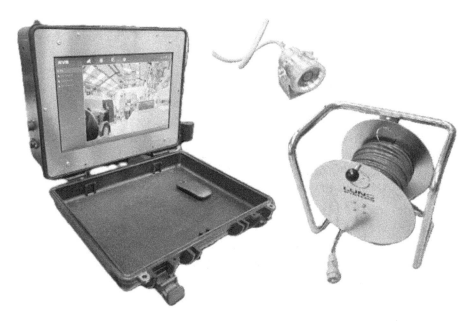

Figure 9.5 Example of a cable reel used for the unbiblical air supply hoses and communication cables to help ensure the hoses and cables don't become tangled or kinked. *Source:* Courtesy of Breathe Safety Ltd. (Luno Systems).

- Each entrant shall have an umbilical cord with a hardened covering to prevent damage. The umbilical cord should contain the primary and secondary air supply hoses, communication cable(s), video cable (although typically not mandatory) and a steel safety cable. There shall be a way to manage the cables, typically a reel, to prevent the cables from getting tangled or kinked. Figure 9.5 illustrates the cables and reel used to maintain them orderly and help prevent getting them tangled or kinked.
- Entrants shall have video capability to help ensure visual contact with the attendants or console operator. This is not required if the line of sight can be maintained between the attendant and the entrants.
- The procedures should specify that switching air cylinders shall not be done when entrants are outfitted in breathing air. Air supply has been lost during switching operations due to freezing condensate or the plugging of air supply lines due to trash in the lines. If the air cylinders run low, the crew must be removed from the inert entry to switch cylinders, which includes a repeat verification of breathing air quality.
- Figure 9.4 illustrates the catalyst technician descending into the reactor with assistance from other catalyst technicians.

9.8 Assuring Breathing Air Quality

Assuring breathing air quality is one of the most important elements of the inert entry procedure. People have died due to mix-ups with breathing air cylinders, cylinders shipped as breathing air but turned out to be something other than breathing air. This generally means certain death for someone who depends on breathing air to sustain life while working in an environment that is immediately dangerous to life and health (the inert reactor entry).

Breathing air cylinders are sourced from suppliers based on their on-site practices for assuring breathing air quality and segregation from potential contamination with other inert gases such as argon or helium. Breathing air for inert entry technicians and others is only supplied from compressed air cylinders certified as breathing air quality. Breathing air supplied by companies or suppliers that are manufactured, synthetic or blended air (air blended by mixing nitrogen and oxygen) shall not be allowed.

Breathing air must meet the breathing air quality standard set by the Occupational Safety and Health Administration (OSHA). Breathing air systems are designed to meet Occupational Safety and Health Administration (OSHA) Standard 29 CFR (1910.134)(i)(1), which states, 'compressed breathing air shall meet at least the requirements for Grade D breathing air described in ANSI/Compressed Gas Association Commodity Specification for Air, G-7.1-1989, to include:

- Oxygen content of 19.5–23.5%
- Hydrocarbon (condensed) content of 5 mg per cubic meter of air or less.
- Carbon monoxide (CO) content of 10 parts per million (ppm) or less.
- Carbon dioxide (CO2) content of 1,000 ppm or less.
- Lack of noticeable odour'.

Or to the European Standards: EN12021:2014:

The quality of compressed air used in breathing apparatus is specified in the European Standard EN12021:2014, 'Respiratory protective devices – Compressed Air for breathing apparatus'. It is vital that the quality of compressed air used in respiratory applications is in accordance with the required standards to prevent risks to human health and corporate reputation.

Each breathing air cylinder delivered to the facility from a breathing air supplier shall be accompanied by a certificate of analysis indicating that the cylinder contains Grade D breathing air that meets the quality outlined above and in the OSHA regulation Standard 29 CFR (1910.134)(i)(1). The certification from the supplier shall state that the cylinder contains pressurized breathing air and is not blended air (a blend of nitrogen and oxygen). The facility shall match the certificate number for each cylinder with the respective cylinder and verify the cylinder is identified correctly and that it is clearly labelled as breathing air.

Each breathing air cylinder delivered to the facility for use in reactor inert entry must be field tested to verify that it meets the OSHA standard for oxygen content before the cylinder is used at the facility. Each cylinder must be tested for percentage of oxygen, and the results must be recorded. The cylinder must be segregated into those that have passed the quality inspection of 19.5–23% from those still pending testing.

Any cylinders not meeting the 19.5–23% requirement should be promptly tagged as 'failed inspection' and immediately removed from the facility. I would expect an investigation to be completed to determine how this occurred. Each cylinder shall be tagged with the test results before being segregated into those that have passed the inspection.

9.9 The Contractor Selection Process

The contractor selection process should review the contractor's work history, especially inert entry jobs. Their incident/accident experience and employee injury statistics, including loss time injury rate and total recordable injury rate, should be considered. Only inert entry trained and experienced personnel should be considered for inert entry roles. The contractor selection process must ensure the contractor selects only highly trained personnel for inert entry assignments. Employee training records and work experience should be available for the facility to review.

9.10 Life Support Equipment

The contractor shall make available the inspection records on all life support equipment to demonstrate that all equipment is carefully maintained and tested. This includes the respirator equipment, including lockable 'clamshell or anti-panic' helmets designed to prevent the wearer from removing the equipment without additional support. As stated earlier, the clamshell helmets and anti-panic helmets incorporate integral breathing apparatus and communications systems.

9.11 Confined Space and Inert Entry Rescue Plan

It is understandable why confined space incidents often involve multiple fatalities. If a worker sees their buddy lying still inside a vessel, it is natural to want to rescue their co-worker. Their first thought likely is that their co-worker has experienced a heart attack or has fallen, and they want to help. However, if a toxic gas has overcome the co-worker, or if the atmosphere has changed due to nitrogen, they too can be overcome. As in the two cases mentioned above, the Valero Texas case and the Canadian refinery case, both co-workers died when attempting to rescue a co-worker. In both cases, they were killed by oxygen deprivation due to a nitrogen-enriched atmosphere. Unfortunately, there have been many more similar events in our industry; nitrogen is truly a silent killer.

Workers MUST be trained; if faced with a co-worker who is down, they must STOP and immediately call for help. They must know that they can be overcome, sometimes with only one breath and they too can die. This is training – workers must know what to do in the event someone is down (call for help) and what not to do (never enter a confined space to rescue unless they are trained as responders and they have the proper breathing apparatus, and then only with a backup person).

According to OSHA 29 CFR 1910.146, employers can provide rescue with in-house personnel or by calling outside emergency services, but either way, the responsibility is to ensure the person is rescued before any long-term harm comes to the person who is down.

The CSB safety bulletin emphasizes that a rescue plan should be developed in advance and implemented during the confined space entry. The plan may include placing the workers in a body harness with a lifeline and wristlets or anklets for workers entering a confined space to accommodate rescue by the attendant in the event of an emergency. Another potential response plan may include having trained response personnel standing near the job site with rescue equipment such as a hoist tripod or similar equipment at the ready. The plan should always include an effective communication plan between personnel inside the space and the attendant and response personnel. Under no circumstances should anyone enter the confined space without proper PPE, including respiratory protection. This includes emergency response personnel.

9.12 Video Surveillance Equipment/Rescue Equipment

The contractor must also provide video surveillance equipment and emergency rescue equipment and adhere to rigorously enforced reactor inert entry procedures.

9.13 Reactor Preparation for Entry

The operators will isolate all feed from the reactor and start a purge of process gas through the catalyst bed to cool the reactor. The process gas should be continued until the reactor catalyst beds are below 38 °F (100 °C). At this point, the reactor is depressurized and drained to remove any residual hydrocarbons from the vessel. The reactor is then isolated from all hydrogen and hydrocarbon sources, and all hydrocarbon or hydrogen lines are blinded with figure-8 or plate blinds to prevent leakage through isolation valves.

At this point, all reactor access/egress points should be posted with warning signs and barricades to ensure all personnel are aware of nitrogen purge and the potential for life-threatening hazards (the creation of a low- oxygen environment near openings where nitrogen may be venting). A personnel access control process should be established to track all personnel entering or leaving the barricaded areas. All personnel entering the barricaded area must be equipped with a personal gas detector, which provides a visual and audible alarm if the oxygen concentration falls below 19.5%. If this occurs, they should leave the area immediately.

The operators will purge the reactor to the flare using the refinery nitrogen header while testing the reactor atmosphere for flammables. Caution: always wear either a hose line-supplied breathing apparatus or a self-contained breathing apparatus (SCBA) when testing the reactor for LEL. Continue purging until the LEL in the reactor is below 10% of the lower explosive limit (LEL).

Once the atmosphere is below 10% LEL, the nitrogen purge should be switched to a dedicated supply (to avoid the risk of contamination), and the purge rate should be reduced to maintain a slight positive pressure in the reactor while the flare connection is positively isolated, and the manway or top elbow, often used as the reactor access point, is removed in preparation for vessel entry. When not in use, the reactor access points should be fully covered to minimize the nitrogen venting and prevent air from being drawn into the reactor. Always be sure to place a temporary hardened cover or grid over the open manway to ensure the cover is reinforced to prevent someone from accidentally falling into the reactor. Remember, this has happened, and people have died as a result.

9.14 Caution

When testing the reactor for 10% LEL, ensure the use of a flammable gas detector designed to measure vapours in an inert atmosphere. Many portable gas detectors used by petroleum refineries and petrochemical plants are not designed for inert conditions and will not provide an accurate LEL in these scenarios. These are typically designed to operate where there is some oxygen present, and the oxygen is used in the performance of the test. To ensure an accurate gas test reading, be sure to use a gas detector specifically designed for use in a scenario where there is no air or oxygen present.

As a reminder, an approved airline-supplied respirator or self-contained breathing apparatus (SCBA) must be worn during gas sampling and at any time workers are near open manholes or vent points during nitrogen purging operations. Remember that temporary enclosures placed over open nozzles or manways for weather protection can inadvertently create an oxygen-deficient atmosphere. Personnel should be aware of this and, if possible, avoid these areas. Workers near the opening of the reactor should also wear fall protection to prevent a fall into the open vessel. SCBA and fall protection when working near the open vessel should be included as a requirement in the procedures.

Refinery standard respiratory protection, including SCBAs, is not suitable for entering and working in an inert entry scenario. As outlined in the equipment discussion above, entry into an inert confined space requires a lockable clamshell or anti-panic type helmet to prevent the wearer from attempting to remove the helmet, for example, in a panic attack. The full-face mask or helmet shall be connected to a full life support system via reinforced umbilical cords designed to prevent damage to the air supply hose lines and communication cables. These systems are designed specifically for use in inert entry into an oxygen-deficient atmosphere.

Before entry, and continuously for all periods entrants are working in the vessel, the atmosphere shall be tested to ensure conditions have not changed. The atmosphere inside the reactor shall be continuously monitored for LEL, oxygen (O_2), hydrogen sulfide (H_2S) and carbon monoxide (CO) using a specialized gas detection system with remote readout. This monitoring shall continue for all periods when entrants are working in the reactor.

9.15 Catalyst Crusting (Another Caution)

Additional special conditions apply for inert entry into multi-bed reactors and catalyst beds susceptible to 'crusting'. Crusting can form on top of the catalyst bed and result in pressure build-up inside the reactor, below the crust. The following shall be followed to verify that crust is not present before allowing a worker to enter a vessel that is being purged from the bottom and vented at the top.

Perform pressure differential testing across the reactor to determine if a crust is present in the reactor. Before the first entry into the reactor, the pressure drop across the reactor should be measured, and the flow of nitrogen to the bottom of the reactor should be stopped momentarily while continuing to watch the pressure at the bottom of the reactor. When the nitrogen flow is stopped, the pressure at the bottom of the reactor should fall quickly. With an open top manway, it should only take a few seconds for the pressure at the bottom of the reactor to suddenly decrease.

If the pressure does not drop immediately, this may indicate that crust may be present on the catalyst. A good indication of reactor crusting is, with nitrogen flowing, there is pressure at the bottom of the reactor while the manway at the top is open. If differential pressure across the reactor is present, personnel shall not enter the vessel with the accumulation of pressure inside the reactor.

Figure 9.6 Catalyst technician fully outfitted and prepared for inert entry. Please note the anti-panic helmet, the harness and the 10-minute escape breathing air pack. *Source:* Courtesy of Breathe Safety Ltd. (Luno Systems).

Figure 9.7 Catalyst technician fully outfitted and prepared for inert entry. Please note the anti-panic helmet, the harness and the 10-minute escape breathing air pack. *Source:* Courtesy of Breathe Safety Ltd. (Luno Systems).

Pressure accumulation below the catalyst is a serious issue; therefore, the crust must be broken so that reintroducing the nitrogen purge does not allow pressure to accumulate. If there is an indication of a crust, the bottom inert gas purge must be shut off temporarily to eliminate the internal pressure. After the pressure has been eliminated, this should allow a worker to enter the vessel and manually break up the crust. After breaking the crust, the bottom purge can be reinstated, continuing to monitor for pressure accumulation in the reactor as work continues.

Figures 9.6 and 9.7 illustrate the catalyst technicians fully outfitted and ready for a catalyst replacement while under inert entry conditions.

End of Chapter 9 Review Quiz

1 How can a catalyst crust be identified in a reactor about to undergo an inert entry to replace the catalyst?

   Answer(s):

2 What is the hazard of a crust on top of the catalyst when planning an inert entry into a reactor?

   Answer(s):

3 What is the significant hazard associated with conducting inert entry into a refining reactor?

   Answer(s):

4 Why is it important to inert the reactor vessel with nitrogen or another inert when opening the reactor to the atmosphere?

   Answer(s):

5 What is the maximum oxygen concentration allowable in the reactor when conducting an inert entry into the reactor? What is the hazard we are addressing with his very low oxygen concentration?

Answer(s)

6 How do we know that the quality of the compressed air meets the OSHA standards for grade D breathing air?

Answer(s):

7 Why is it important to have a preplanned rescue plan and trained responders on site during the vessel entry?

Answer(s):

8 Why are specialized clamshell or anti-panic helmets required for entrants when conducting inert entry?

Answer(s):

9 What is the maximum LEL (flammables) allowable before entry into a reactor vessel when undergoing inert entry for catalyst replacement?

Answer(s): Please select the most correct answer.

A 2%
B 5%
C 10%
D 15%
E 25%

10 When the test for carbon monoxide is positive, what chemical may be present in the reactor?

Answer(s):

11 During an inert entry, how many breathing air sources are available to those working in the reactor?

Can you name the breathing air systems? Answer(s):

12 What special PPE should be worn by the workers who are testing the reactor for LEL before permitting entry into the inert atmosphere?

Answer(s):

13 The primary nitrogen supply during vessel entry should be a _____ system and from a reliable source. A backup nitrogen supply that is completely _____ of the primary supply should be readily available and also reliable. Entry is only allowed when using the _____ source.

Answer(s):

14 What is the maximum oxygen allowed in the nitrogen purging system?

Answer(s):

15 Before the entry, the inert entry contractor must submit a detailed _____ _____ _____ to the facility for approval/endorsement.

Answer(s):

16 At most companies, are company employees allowed to suit up and enter a vessel with an inert atmosphere?

Answer(s):

17 During most inert activity tasks, who is monitoring the breathing air pressures and breathing air rates and how is it monitored?

Answer(s):

Additional Resources

API Standards 2217A, Guidelines for Safe Work in Inert Confined Spaces in the Petroleum and Petrochemical Industries, Fourth Edition, July 2009. Luno Systems, Guidance on Safe Entry to Inert Confined Spaces. https://www.lunosystems.com/files/GuidanceonSafeEntrytoInertConfinedSpaces InertEntry.pdf

European Catalyst Manufacturers Association Avenue Edmond van Nieuwenhuyse 4, 1160 Brussels, Belgium, Tel. +32 2 676 72 24, bpi@cefic.be. Catalyst Handling Best Practice Guide. https://www.catalystseurope.org/images/Documents/ECMA1004q_-_Catalyst_handling_best_practice_guide.pdf

Engineers Ireland, Catalyst handling - mitigating hazards relating to inert entry. https://www.engineersireland.ie/Brexit/catalyst-handling-mitigating-hazards-relating-to-inert-entry

British Health and Safety Executive, Inerting. https://www.hse.gov.uk/comah/sragtech/techmeasinerting.htm

US Department of Energy, Office of Health Safety and Security, Just-in-Time Report. https://ehss.energy.gov/sesa/Files/Analysis/jit/JIT_2006-04.pdf

Digital Refining, Worker safety when entering reactors with an inert atmosphere (TiA). https://www.digitalrefining.com/article/1003058/worker-safety-when-entering-reactors-with-an-inert-atmosphere-ti

10

Carbon Capture, Use, and Storage

This chapter discusses the technology to remove carbon dioxide (CO_2) from the environment and prevent it from contributing to greenhouse gas emissions. The Carbon Capture Storage (CCS) and the Carbon Capture, Utilization and Storage (CCUS) technologies are currently under development. These two technologies are very similar. The only significant difference is that CCS involves only CO_2 capture and storage, while CCUS also involves the potential beneficial uses of CO_2.

This project is primarily an environmental strategy, so why am I concerned about it as it relates to inert gas? The goal of this project is to capture carbon dioxide from the atmosphere and ongoing operations, as well as store or 'sequester' the recovered gas to prevent it from contributing to greenhouse gases, which can harm the environment. This means that existing technology may be modified, or new technology may be developed and implemented to capture and store the CO_2.

We have discussed CO_2 as an inert gas and that it is an asphyxiant, especially in high concentrations or in enclosed areas or areas where there is little or no air circulation. People have been killed by exposure to CO_2 due to the displacement of oxygen from their environment.

These hazards must be considered during the design, construction, start-up and online operational phases of the modified or new equipment being brought online. All designs must consider the hazards of CO_2 and ensure that personnel are protected from these unique hazards. Our existing project management systems and procedures, such as Management of Change, Hazard Operability Study (HAZOP), Pre-start-up Safety Review (PSSR), Operating Procedures, etc., should be adequate, but they must be fully utilized during all phases of the modified or new projects. This is true whether the facility falls under OSHA Process Safety Management or not.

Carbon dioxide (CO_2) occurs naturally in the atmosphere from off-gassing from the ocean, decaying vegetation, wildfires and even volcanos. However, most of the CO_2 is produced from the burning of fossil fuels, such as coal, oil, natural gas, waste and biological materials (e.g. trees). Some chemical reactions also result in the generation of CO_2. There is also a natural removal process of carbon dioxide from the atmosphere. This happens because the trees and other plants on earth take in CO_2 through a process of 'sequestration' and return oxygen to the planet.

Unfortunately, carbon dioxide (CO_2) is a significant contributor to the greenhouse gas that plays a primary role in global warming of the planet. Much of that greenhouse gas is believed to come from human activities. According to the Environmental Protection Agency (EPA), in 2022, CO_2 accounted for 80% of all US greenhouse gas emissions from human activities. The natural carbon cycle is being changed by human activities. This works in two ways: humans are adding more CO_2 to our normal day-to-day activities, and we are impacting the ability of forests and other natural sinks to remove the CO_2 and store it away from the atmosphere.

Hazards of Nitrogen and Other Inert Gases: How They Can be Safely Managed, First Edition. M. Darryl Yoes.
© 2025 John Wiley & Sons, Inc. Published 2025 by John Wiley & Sons, Inc.

An interesting fact about CO_2 is that at normal room temperature and pressure, CO_2 is a gas and is not naturally present in liquefied form. At low temperatures, it exists as 'dry ice'. When the temperature rises above −109 °F (197.4 °K [Kelvin] or −78.45 °C), the solid CO_2 (dry ice) sublimes to the gaseous form of CO_2. However, CO_2 can be liquefied, but only under specific conditions of high pressure and low temperature. It is a liquid state when compressed and cooled to below its critical point.

Note: there are dissenting voices to the whole carbon capture initiative. This is clear, and to share this, I have posted several links to these sites in the Additional Resources section of this chapter. One clear thing is that the US Government is pouring many resources, including what appears to be billions of dollars, into this, and many companies have lined up with projects to address carbon capture. I am not one to say, but what I am attempting to do is to ensure that if we have projects moving forward in your company, I want to do what I can to ensure that they are being constructed and operated with safety in mind. This means both personnel safety and operations safety.

Many risks are associated with carbon dioxide facilities and operations, and these must be recognized and addressed. We certainly have the skill, talent and know-how to make this happen safely.

10.1 How is the Industry Responding to This Increase in Carbon Dioxide?

CCUS is considered an option to help reduce gas emissions that contribute to greenhouse gas formation in the United States and worldwide. Therefore, CCUS is considered a key approach in the current range of options available. Carbon capture and storage of CO_2 have been identified as cost-effective ways to permanently remove CO_2 from the environment and decarbonize emission-intensive industries. Two international groups, the United Nations Intergovernmental Panel on Climate Change and the International Energy Agency, have helped provide focus to help promote this within the industry. The Center for Climate and Energy Solutions has said that they expect 90% of CO_2 emissions from electrical power plants and other industries, such as petroleum and petrochemicals, iron and steel manufacturing and cement production, can be captured, especially when combined with bioenergy or direct air capture technologies.

There have been several potential alternative uses for the recovered carbon dioxide. One of the most promising is the use of CO_2 in the enhanced recovery of hydrocarbons from underground wells. This extraction method uses CO_2 to drive the oil or gas from the wells and sequesters the CO_2 underground in the depleted wells. Another option is selling CO_2 as a product for use in well sites to enhance oil recovery. This could prove profitable for facilities and a financial incentive to pursue additional CCUS technologies.

The Intergovernmental Panel on Climate Change (IPCC) recently highlighted that we must do more than simply reduce emissions to meet the goals of the Paris Agreement and thereby limit future global temperature increases. They stated that we also need technologies that can remove CO_2 from the atmosphere. Several of these technologies are discussed below.

1. Carbon Capture
The CO_2 is separated from gas or waste streams that are produced in industrial processes, such as coal or natural gas-fired power generation, cement manufacturing, refining or petrochemical

plants, iron or steel manufacturing facilities and others identified as emission-intensive facilities. A suite of energy technologies is designed to capture carbon dioxide (CO_2) from these emission-intensive facilities. These technologies are referred to as Carbon Capture Usage and Storage (CCUS). This suite includes several technologies that are proven to remove other contaminants or emissions from other process streams and can be adapted or used for CO_2. There are several others that appear feasible but are not yet proven. The CCUS technologies include the following.

Carbon Capture is an emerging technology, and at this time, it involves primarily three strategies:

- Post-combustion carbon capture: primarily used in existing power generation facilities.
- Pre-combustion carbon capture: primarily used in industrial processes like refining and petrochemicals.
- Oxy-fuel combustion systems: a relatively new technology with one pilot plant in progress. Several of these processes are listed next. This is not intended to be a complete list.
  - Chemical Solvents (Solvent Extraction):
  Amine scrubbing is one of the most promising chemical solvents; however, all solvents work basically the same. The solvent works by the solvent contacting the CO_2 in a process stream or effluent stream. The solvent attracts the CO_2, and the CO_2 attaches to the solvent. This is typically referred to as a 'rich solvent'. Afterward, the solvent is processed in a separate separation vessel where it is heated using steam or another medium to release the CO_2 for capture. The 'regenerated solvent' is recycled back to the scrubber, where it is reused in the process. Solvent extraction of this type is a proven technology and is frequently used in refineries and petrochemical plants to remove hydrogen sulphide and/or other contaminants from process streams.

  One example of a chemical solvent that has proven successful for CO_2 removal is in use at the Petra Nova power plant in the Houston area. This is the first commercial-scale carbon capture facility at a US power plant. This facility uses a solvent that has been under development since the early 1990s. This site reportedly captures SO_2 emissions roughly equivalent to about 350,000 vehicles.

  Treatment with chemical solvents requires significant investment in process equipment, such as gas scrubbers to scrub the process gases and stripping facilities to remove the CO_2 from the solvent and recover the solvent stream for reuse.
  - Sorbents:
  Sorbents work similarly to solvents, except in the case of sorbents, the CO_2 is adsorbed onto the solid sorbent. Like with solvents, the CO_2 must be removed from the sorbent, which is typically done by heating the solid sorbent or by reducing the operating pressure on the sorbent. This promising technology is still in the design/experimental stage and has not been proven when compared to the solvent technology.
  - Membranes:
  Carbon dioxide is captured in membrane systems by flowing the gas containing the CO_2 through the membrane cells. These function like a strainer where the smaller molecule gas flows readily through the strainer, but the larger gases will not flow through the membrane. Membrane performance is defined by the rate at which CO_2 can flow through the membrane. We specify the membrane functions as a combination of permeance (the flow rate of the selective molecule, in this case, CO_2) and selectivity (the degree to which one molecule gets through vs. others). Of course, our preference is to have membranes that have both high permeance and high selectivity, but these tend to offset each other. However, membranes appear to have a role in CO_2 capture, although more work needs to be done in this important area.

- Cryogenic Systems (Cryogenic Fractionation):
  Cryogenic fractionation of air to produce oxygen, nitrogen, argon and other gases is nothing new and has been around for decades. This process supercools the air until the fractions are liquefied, and the various fractions are then separated by boiling points to form purified gas streams. However, this process is very energy-intensive, requiring multiple stages of compression, refrigeration and expansion. This requires lots of energy, typically electricity or high-pressure steam-driven compressors.

  One approach under consideration for CO_2 separation is to cool the CO_2 until it becomes dry ice, a solid. Therefore, either solid handling facilities will be required, or dry ice will be allowed to sublime to form CO_2 gas. Sublimation occurs at a temperature of $-109.3\ °F\ (-78.5\ °C)$.

  Research continues on this technology, and progress is being made.

- Oxy Combustion:
  The fuel is combusted with essentially pure oxygen instead of air in these sophisticated systems. Air has 21% oxygen, compared to firing these systems with essentially 100% oxygen. The waste stream from an oxy combustion system is essentially carbon dioxide and water, which makes the separation of CO_2 easy. The only critical separation process is the separation of oxygen from the air to allow the use of the oxygen to combust the fuel.

  An oxy combustion system is being tested in a pilot plant in the Houston area. This project expects to create a zero-emission-based natural gas power plant. This technology fires natural gas with pure oxygen, and the resulting gas drives a specially designed turbine for power generation. The turbine exhaust is cooled to remove water and recycle heat, leaving a stream of nearly pure CO_2.

2. Carbon Transport

The CO_2 is then compressed and transported via pipelines, road transport or ships to a site for storage.

The most common and by far the most efficient and safest way to transport carbon dioxide is by pipeline. The principle behind transport by pipeline is that liquid or gas moves from an area of higher pressure to an area of lower pressure.

There are many gas and oil pipelines worldwide, including pipelines that are used to transport carbon dioxide. In the United States alone, there are over 50 CO_2 pipelines, including approximately 4,000 miles (6,500 km).

> Compressed gas requires less volume when it's compressed and far less when it is liquefied or solidified. Therefore, in preparation for transport, CO_2 is frequently chilled and compressed to a liquid state before transport. It is frequently transported as a supercritical fluid.
>
> CO_2, when in the supercritical state, is about the same density as a liquid or mass per volume (how tightly a material is packed together). However, the viscosity, or thickness, is similar to gas. The lower the viscosity of a liquid, the thinner it is and the lesser the resistance to flow. This makes supercritical liquids much easier to pump through pipelines.

This means that CO_2, when in the supercritical state, requires much less space on-board ships and can be loaded in greater quantities. Also, supercritical liquids generate much less friction when moving through pipelines, requiring less energy and moving at higher rates.

Pipelines represent a safe way to transport CO_2. However, there are a number of challenges ahead, and the design and construction of CO_2 Pipelines will be expensive and take time. The pipelines must be specially designed to transport low-temperature and high-pressure CO_2. The low temperatures result in the piping becoming brittle and must be fabricated using special metallurgy and will require frequent and expensive inspections. Existing oil and gas pipelines will not be suitable. CO_2 will have to be essentially pure and free from water and other impurities that can corrode or otherwise damage the pipelines. Otherwise, the risk is loss of containment and explosions from rapidly expanding CO_2, as the liquid turns to gas due to a pipe failure and boiling liquid expanding vapour explosion. The pipeline must also be connected to the final storage cavern or destination, which may be long distances from the recovery facility.

When CO_2 is transported on ships, it is cooled to a very low temperature and compressed to change state to a liquid (liquefied carbon dioxide) (CO_2), similar to the way natural gas is liquefied for shipment. When liquefied, the CO_2 is loaded into specially designed cargo tanks on board the vessel. These cargo tanks are insulated, designed and constructed to withstand low temperatures and higher pressures. This facilitates efficient and safe transportation over great distances.

3. Carbon Storage
There are currently examples where storage has been identified, and carbon capture, transport and storage are either in place or where projects are well underway. For storage, ideally, the CO_2 will be injected into rock formations deep underground for permanent storage, typically at least 1 km underground. Other possible storage sites may include saline aquifers and/or oil and gas reservoirs or coal beds that have already been depleted of production. Some of these sites are in the United Kingdom and the North Sea. Other possible storage sites for CO_2 emissions include saline aquifers or depleted oil and gas reservoirs, which typically need to be 0.62 miles (1 km) or more under the ground. For example, a saline aquifer in the North Sea, approximately 56 miles offshore (90 km) and about 1 mile below the sea, is being considered to store very large amounts of CO_2. Likewise, another similar site in Alabama (United States) is also under consideration.

10.2 Safety Aspects of Carbon Capture, Use, and Storage (CCUS)

In this section, we will only talk about the safety aspects of CCUS. This is a very broad topic, and therefore, we will only cover this topic from a 10,000-foot elevation with little detail provided. However, this may give the reader a better understanding of the challenges and some approaches that may be used as we go down this path.

10.2.1 Design and Construction

Due to the processing of CO_2, which is an inert gas with the potential for asphyxiation, this should be considered in all aspects of equipment design. Equipment should be designed where isolation is straightforward and can be quickly executed, and the equipment can be vented to a safe location. Also, some equipment will be designed to process liquefied CO_2 at extremely low temperatures and high pressures. In most cases, this will require specialized equipment made of alloy materials. Consideration should be included for the equipment expansion/contraction as temperatures rapidly change during start-up and shutdown. Adequate tolerances should be included for equipment corrosion and designed to facilitate online inspection and repairs. Piping design should not include dead legs or dead ends (sections with little or no flow), and all small connections should be seal welded and protected with gussets (braces).

All equipment in these services must fully comply with applicable design standards, such as API, ASME, ASTM, ANSI, NFPA, NPRA and AWS. Design in other countries uses many of these same standards, plus they may also have standards of their own, such as EN in Europe (European Standards).

A Process Hazard Analysis (PHA) should be completed during the various stages of design to ensure safety issues are identified and addressed. The facility should ensure that a Pre-Start-up Process Hazard Analysis is completed, and all outstanding recommendations are completed or otherwise addressed prior to preparation for the start-up of the new or modified facilities.

A detailed Management of Change document should also be completed, including a Pre-start-up Safety Audit before the new or modified equipment is placed in service. Please refer to CFR 29 CFR 1910.119 (for petroleum refining or petrochemical sites in the United States) or the specific regulatory requirements for your facility.

10.2.2 Start-up and Operation

Start-up and Operations should be prepared in advance with a detailed set of Start-up Procedures, including Sections for Normal Operations, Temporary Operations, Emergency Operations, Normal Shutdown, Start-up and follow the requirements set out in OSHA 29 CFR 1910.119 for petroleum refining and petrochemical facilities or the appropriate regulatory requirements for your facility. Operating procedures should be in a checklist format with signoffs for each step by the Operator. Notes or other guidance should be kept to a minimum, and steps should not be embedded in notes.

All work in or near process equipment with a possibility of CO_2 release should require the use of an oxygen detector and a fresh-air-supplied respirator, such as a self-contained breathing apparatus (SCBA).

As stated above, it is imperative that a Pre-start audit be completed and that all identified issues be either resolved or addressed to help ensure a smooth and safe start-up.

10.2.3 Transport

The transport medium, whether a truck, a railcar, a ship, a pipeline or some other means, must have equipment specially designed to handle the low temperatures and high pressures associated with liquefied carbon dioxide. This means the transport piping or vessel must be designed and constructed to handle the potential embrittlement concerns and the high pressures involved. Equipment such as pipelines should be designed for Operators to detect system leakage and to have the ability to safely and quickly isolate and depressurize the system in the event of a loss of containment incident.

All work in or near process equipment with a possibility of CO_2 release should require the use of an oxygen detector and a fresh air-supplied respirator, such as a self-contained breathing apparatus (SCBA).

10.2.4 Storage or Sequestration

Storage of CO_2 should be in deep rock crevices or formations such as depleted oil and gas wells, coal beds, saline aquifers or similar caverns or structures. There are several considerations for this storage, such as potential leakage to the atmosphere, the potential for rupture of the underground structure leading to something like tectonic movement (earthquake or similar), the potential formation of sinkholes, etc. These should be addressed with a thorough geological survey, typically completed by a hydrologist, geologist or geotechnical engineer, to ensure that the formation can contain the stored CO_2 without leakage or geological failure.

Abbreviations

API (American Petroleum Institute Codes)
ASME (Boiler and Pressure Vessel Codes)
ASME Piping Codes
ASTM (American Society for Testing and Materials)
ANSI (American National Standards Institute)
AWS (American Welding Society)
NFPA (National Fire Protection Association)
NPRA (National Petroleum Refiners Association)
EN (European Standards)

End of Chapter 10 Review Quiz

1 Why is carbon capture so important?
   Answer(s):

2 Which inert gas is carbon capture focused on to remove it from our environment and help prevent the formation of greenhouse gases?
   Answer(s):

3 CO_2 is an inert gas, and it is an asphyxiant, especially in high concentrations or in enclosed areas or areas where there is little or no air circulation. People have been killed by exposure to CO_2 due to the displacement of _____ from their environment.
   Answer(s):

4 When building new projects or modifying others to extract and process CO_2, the hazards must be considered during the _____, _____, _____ and _____ operational phases of the modified or new equipment being brought online.
   Answer(s):

5 CO_2 does not naturally exist in a liquid state. However, it can be liquefied, but only under specific conditions of high _____ and low _____.
   Answer(s):

6 According to the Environmental Protection Agency (EPA), in 2022, CO_2 accounted for _____% of all US greenhouse gas emissions from human activities.
   Answer(s):
   A 10
   B 25
   C 65
   D 80
   E 90

7   What is a leading alternative use of carbon dioxide after it is captured from the environment?

    Answer(s):

8   What are the five current processing strategies or technologies being explored for carbon capture?

    Answer(s):

9   The most common and by far the most efficient and safest way to transport carbon dioxide is by _____.

    Answer(s):

10 What is meant by the term sequestration?

    Answer(s):

Additional Resources

US Environmental Protection Agency (EPA), Greenhouse Gas Emissions. https://www.epa.gov/ghgemissions/overview-greenhouse-gases#:~:text=Global%20Warming%20Potential%20(100%2Dyear,gas%20emissions%20from%20human%20activities.

US Environmental Protection Agency (EPA), Carbon Dioxide Capture and Sequestration: Overview. https://19january2017snapshot.epa.gov/climatechange/carbon-dioxide-capture-and-sequestration-overview_.html

US Department of Energy (DOE), DOE Explains…Carbon Sequestration. https://www.energy.gov/science/doe-explainscarbon-sequestration

US Department of the Interior (DOI), US Geological Survey, What is carbon sequestration? https://www.usgs.gov/faqs/what-carbon-sequestration

International Energy Agency (IEA), Carbon Capture, Utilisation and Storage. https://www.iea.org/energy-system/carbon-capture-utilisation-and-storage

Wikipedia – The Free Encyclopedia, Carbon Dioxide. https://en.wikipedia.org/wiki/Carbon_dioxide

ExxonMobil, Carbon Capture and Storage. https://corporate.exxonmobil.com/what-we-do/delivering-industrial-solutions/carbon-capture-and-storage?utm_source=google&utm_medium=cpc&utm_campaign=1ECX_GAD_TRAF_OT_Non-Brand_Carbon+Capture&utm_content=OT_Non-Brand_Carbon+Emissions&utm_term=carbon+emissions&gad_source=1&gclid=EAIaIQobChMIxJ215OapiAMVvUL_AR1rcyPHEAAYAyAAEgJYP_D_BwE&gclsrc=aw.ds

Chevron Explainer: what is a CO_2 storage hub? https://www.chevron.com/newsroom/2023/q2/explainer-what-is-a-carbon-storage-hub?utm_source=GGL&utm_medium=cpc&utm_campaign=Chevron_National_Nonbrand_Explainers_Articles_Exact&gad_source=1&gclid=EAIaIQobChMIo5_D6sy0iAMVSkt_AB1NPRXKEAAYBCAAEgKvxvD_BwE&gclsrc=aw.ds

Carbon Capture and Storage 101, Resources for the Future – Resources Magazine. https://www.rff.org/publications/explainers/carbon-capture-and-storage-101/?gad_source=1&gclid=EAIaIQobChMIzr6Z4P2riAMVeDjUAR34iAWTEAAYAyAAEgJKyvD_BwE

What is the Carbon Cycle? Drax Group - A renewable energy company. https://www.drax.com/carbon-capture/what-is-the-carbon-cycle/

Carbon Capture – Opposing Views

Food and Water Watch, The Scam of Carbon Capture – Carbon Capture is Dangerous Carbon Capture: Billions of Federal Dollars Poured Into Failure. https://www.foodandwaterwatch.org/2022/09/27/carbon-capture-failures/?gad_source=1&gclid=EAIaIQobChMIhbWsw62siAMVrjrUAR07rw5gEAAYASAAEgKy7fD_BwE

The Understory, Public Interest Groups Oppose Carbon Capture Scam. https://www.ran.org/the-understory/public_interest_groups_oppose_carbon_capture_scam/?gad_source=1&gclid=EAIaIQobChMIhbWsw62siAMVrjrUAR07rw5gEAAYAiAAEgLSxvD_BwE

Dissent Magazine, The Carbon Capture Distraction. https://www.dissentmagazine.org/article/the-carbon-capture-distraction/#:~:text=Critics%20often%20argue%20that%20CCS,of%20electricity%20as%20one%20without

11

Nitrogen Asphyxiation Case Studies, and Asphyxiation by Other Inert Gases

What follows next are case studies of nitrogen asphyxiation incidents that occurred in the processed food and petroleum refining industries. The first two industries don't have a lot in common. However, they both use nitrogen and sometimes liquid nitrogen to purge equipment and for the super cold properties of liquid nitrogen. These two incidents resulted in the loss of the lives of eight workers with far-reaching impacts on the families of these workers. Both incidents were investigated by the US Chemical Safety and Hazard Investigation Board (CSB) and by the US Occupational Safety and Health Administration (OSHA). Detailed reports are available for both incidents, and they are summarized in this book. More information on these incidents is available in references provided in the Additional Resources section.

Also, an Appendix titled 'Documented Incidents Involving Nitrogen Resulting in Fatalities or Serious Injury' is available, detailing numerous other cases of loss of life and serious injuries resulting in asphyxiation due to nitrogen exposure. I have over 50 years of experience in the refining industry, and I was aware of several of the higher profile incidents that have occurred. However, I was absolutely astounded when I pulled this list of incidents together at the number of incidents and the number of lives lost due to nitrogen asphyxiation. I was even more astounded at the number of incidents that almost read the same or repeats of the previous incidents that occurred, resulting in loss of life or serious injury. We don't seem to learn or communicate what has happened before, and the same incidents are repeated. This list of incidents was pulled from several databases, primarily the OSHA and CSB databases.

I have categorized these reports into the applicable causal factors and discussed these as separate groups. However, never forget that the common cause of all of these incidents was directly related to the use of nitrogen by workers doing what they felt was safe. In each case, the individual workers failed to recognize the hazards of nitrogen. In multiple ways, they placed themselves and others directly in harm's way in an inert but extremely dangerous environment, where they were either killed or at grave risk of being killed.

The complete list of OSHA reports is available in the Appendix.

I hope you will take the time to look through the information provided in Appendix 3 and become familiar with the kinds of incidents that have occurred in the past and are still happening today. As you read through the summarized incident reports in the Appendix, please remember that these guys and gals were sons and daughters, brothers and sisters, and teammates. None of these people came to work on the morning of their incident, thinking it to be the day that they would die. These are truly tragic incidents when something like this happens. A quick summary of the information from the Appendix is provided in Table 11.1. Hopefully, you will take the information provided in the Appendix and in this book back to your workplace and ensure that there are

Hazards of Nitrogen and Other Inert Gases: How They Can be Safely Managed, First Edition. M. Darryl Yoes.
© 2025 John Wiley & Sons, Inc. Published 2025 by John Wiley & Sons, Inc.

Table 11.1 Review of Appendix: numbers of documented incidents involving nitrogen resulting in fatalities or serious injury in the United States.

Number of incidents	Number of deaths	Number of injuries	Scenario resulting in asphyxiation or potential asphyxiation
26	26	2	The number of incidents where the worker(s) connected their respirator or sandblast hood to a facility pipeline in nitrogen service or a facility pipeline normally in air service but backed up by nitrogen at the time. An incident of this type most often results in the death of the worker, and most of these incidents result in at least one fatality.
70	69	16	The number of incidents where the worker(s) enters a tank, truck, or other process vessel without verification that the atmosphere would support human life, without an entry permit, or without a fresh air-supplied respirator. Almost all these incidents also resulted in the death of the worker(s).
14	11	6	The number of incidents where an untrained and non-equipped person attempted to rescue someone else. Generally, this also results in the person attempting rescue also being killed.
5	5	0	The number of incidents in which the worker(s) were near an opening being purged with nitrogen with no intent to enter. Many of these incidents resulted in the person being overcome and falling into the opening or vessel.
2	5	0	The number of incidents when a breathing cylinder containing a gas other than breathing air was used (total of five deaths, including four nursing home patients who died due to breathing nitrogen).
4	4	0	The number of incidents that occurred due to the worker(s) in an immediately dangerous to life and health (IDLH) environment running out of breathing air in their self-contained breathing apparatus (SCBA).
4	4	1	The number of incidents while the workers were working in an IDLH atmosphere with the wrong type of respirator (organic filter respirator instead of breathing air).
8	8	2	Failure of nitrogen containment and a release in an enclosed space (due to piping or equipment leaks).
1	2	0	Transporting nitrogen cylinders in an enclosed vehicle.
134	134	27	Subtotals for the period March 1984–May 2023.

ironclad procedures and training in place to help prevent an incident like this from happening at your facility. After all, that is exactly what this book is about.

11.1 What Has Happened / What Can Happen Incident Case Study Number 1 – Nitrogen Asphyxiation

Fatal Liquid Nitrogen Release at Foundation Food Group (Six Fatalities, Four Serious Injuries)

Gainesville, Georgia

Incident Date: 28 January 2021

The US Chemical Safety and Hazard Investigation Board investigated this incident and issued a detailed investigation report. A summary follows: please see the CSB full report for more information.

The CSB report provides additional insight into the extreme risks that working with or near nitrogen poses to the lives of workers. This is true even for industries and facilities where nitrogen has been used for decades.

Similar to other poultry and meatpacking plants, the Foundation Food Group facility in Gainesville, Georgia, used liquid nitrogen to instantly freeze its poultry products. Unfortunately, the hazards of working with liquid nitrogen were not recognized at Foundation Food Group until this deadly incident.

There are significant industry cautions and warnings about the use of nitrogen and the important safety precautions that should be in place, including the requirement for atmospheric monitoring, alarms and warnings, and adequate emergency training and preparedness in facilities using liquid nitrogen. The US Chemical Safety Board has been a significant advocate in this important process safety area. However, the CSB found that since there were no warnings, no training and no specialized protective equipment, workers could not recognize the presence of deadly nitrogen gas. Nitrogen in either the liquid or gaseous state is odourless and colourless, presenting no warning that it is present in deadly concentrations.

It is also painfully obvious that Foundation Food Group had not adequately warned workers about the deadly gas's presence. The CSB reported that the workers had received no training about working with and near nitrogen, and there were no alarms that nitrogen was present in deadly concentrations. When the release occurred, 14 employees entered the area, which was contaminated with high concentrations of nitrogen, attempting to rescue or render assistance to the first two employees who had already succumbed to the nitrogen. These 14 were not trained to rescue and had no personal protective equipment, such as self-Contained breathing Apparatus (SCBA) and no personal gas detection monitor.

The CSB determined that the release occurred in an immersion freezer, a device that is designed to flash-freeze cooked poultry products using the auto refrigeration properties of liquid nitrogen (see Figure 11.1 for an image of the immersion freezer). The CSB also found that Foundation Food

Figure 11.1 Immersion freezer at the Foundation Food Group. *Source:* US Chemical Safety and Hazard Investigation Board/Public domain.

Group had placed the immersion freezer in a work area with very little natural ventilation and no forced ventilation. Ventilation is important to help ensure that any nitrogen released by the equipment is quickly purged to ensure that it is safe for the workers in the work areas.

Investigators discovered that the immersion freezer was designed with a single-point liquid nitrogen level control system described as a 'bubbler tube' used to measure liquid nitrogen levels inside the plant's freezer. The bubbler tube works by maintaining a constant flow rate of vapour through the tube and into the liquid and measuring the differential pressure between the liquid and atmospheric pressure.

The immersion freezer control system then used the differential pressure input measurement to calculate the liquid level in the bath and automatically adjusted the position of the control valve, controlling the liquid nitrogen flow to maintain the liquid level at the user-specified set point.

The immersion freezer was also equipped with a high-level safety interlock that, when activated, closed valves to shut off the flow of additional liquid nitrogen into the freezer, thereby preventing an overflow. However, the level control loop and the high-level safety interlock both used the bubbler tube as the input sensor, making this a single point of failure. There was no backup system or device to detect a high level or stop liquid nitrogen flow into the freezer.

Figure 11.2 illustrates the bubbler tube as it was presented to the CSB with the tail of the tube bent in a 90° position. The bubbler tube was designed to vent straight down to measure the differential pressure. The CSB report indicated that the tube may have been damaged during maintenance operations. The bent bubbler tube read a false differential pressure and allowed liquid nitrogen to overflow from the freezer, which quickly filled the room with nitrogen.

There was no high-level alarm, typically a device designed to sound audible or visible alarm, indicating that the level was too high. The liquid nitrogen, when released, quickly vapourized,

Figure 11.2 The immersion freezer bubbler tube (level control device). Notice the thick layer of ice on the floor due to the released super cold liquid nitrogen. *Source:* US Chemical Safety and Hazard Investigation Board/Public domain.

expanding the volume about 700 times, quickly filling the room with the vapourized nitrogen. Nitrogen has no odour. However, under the right climatic conditions, it will create a cold white fog, but it is otherwise invisible.

Note that the CSB determined that this liquid measurement device had been bent approximately to 90°. This device provides liquid measurement for both the level controller and the automatic shutdown system, designed to stop the incoming flow of liquid nitrogen in the event of a high level in the freezer. Both safety devices malfunctioned due to this device failure.

The vapourized nitrogen quickly formed a white cloud of about 4–5 feet and quickly killed two of the employees. As stated earlier, 14 other employees either entered the room where the release occurred or ventured very close as they investigated the release and attempted to provide aid and assistance to the first two victims. Four other employees died from the nitrogen exposure, and three workers and a firefighter who responded to the incident also received serious injuries, as reported by the CSB.

In scenarios like this, where large quantities of nitrogen or other toxic gases or asphyxiants, such as nitrogen, which can harm workers, are involved in the process at the facility, analysers are typically continuously monitoring the atmospheric conditions in the work areas, especially confined spaces such as processing rooms. The report said the plant had also failed to install air monitoring and alarm devices that could have alerted workers about the dangerous vapour cloud and prevented them from entering the freezer room. In a case like nitrogen, the analysers normally monitor the oxygen concentration and alarm at around 19.5–20% O_2 in the work area. This would warn the workers to evacuate the area immediately and ensure adequate ventilation.

The CSB provided several important recommendations in this report, which, when implemented, should help prevent future incidents. CSB Chairperson Steve Owens was quoted as saying, 'The hazards of liquid nitrogen must be clearly communicated to workers, and the safety management systems for operations that use liquid nitrogen must be improved'.

The Chemical Safety Board also advocated for the US Occupational Safety and Health Administration (OSHA) to create a national standard specifically covering the use of liquid nitrogen for the poultry processing and food manufacturing industry.

Key Lessons Learnt

- **Nitrogen is an extreme hazard:**

  Nitrogen is an extreme inhalation and asphyxiating hazard. It is present in high concentrations (78%) in the air we breathe. It works by displacing the oxygen in the air we breathe; when nitrogen concentration increases, oxygen concentration decreases.

- **Process design and process hazard analysis:**

  It is important that the process design for new or modified equipment is reviewed for inherent process hazards. All areas with the potential for loss of containment should be considered for both a primary and secondary control system designed to prevent loss of containment. These instrumented control systems should be completely independent of each other, with no single point of detection or control.

- **Management of change:**

    All changes or modifications made to existing equipment require a written and approved management of change. This includes a review of the potential of introduced process safety hazards and training of employees to be completed before the change is operated. This includes changes made to equipment during maintenance. For example, changing or bending the bubbler tube.

- **Employee training and hazard communication:**

    Employees and others operating the process should be trained in the properties of the material being used in the process. Not all employees fully understand the properties of nitrogen and may expose themselves and others to an environment that would not support life.

- **Emergency drills and exercises:**

    Employees and others working in areas where toxics or asphyxiants are used in the process should participate in periodic emergency response drills or exercises. This is important for workers to understand what actions to take and how and where to respond in the event of a loss of containment.

- **Alarms and warning devices:**

    Facilities using toxics or asphyxiants in the process should have adequate alarms and warning devices to ensure that employees and others are adequately warned of the presence of these materials in the workplace. These devices must be tested periodically to ensure they are maintained in working order. The fixed alarms can and should be augmented by portable alarms or personal gas detection monitors worn by the workers. When properly instrumented, fixed alarms can also be programmed to stop the flow of chemicals. Workers should be trained to immediately leave the area when/if the alarms are active.

- **Emergency procedures:**

    Emergency procedures should be adequate for workers to understand potential scenarios, written in the language(s) used by the workers in a step-by-step format. Employees should receive training in the initial and refresher procedures, and procedures should be readily available to all workers.

- **Ventilation in confined spaces:**

    Ensure that confined rooms or spaces where toxics or asphyxiants such as nitrogen are used or processed and where humans are present are adequately ventilated or equipped with forced ventilation.

11.2 What Has Happened / What Can Happen Incident Case Study Number 2 – Nitrogen Asphyxiation

Fatal Nitrogen Exposure at Valero Delaware City Refinery (Two Fatalities)
Delaware City, Delaware

Incident Date: 5 November 2005

The US Chemical Safety and Hazard Investigation Board also investigated this incident and issued a detailed investigation report. A summary follows: please see the CSB full report for more information.

On 5 November 2002, two contract employees working at the Valero Energy Corporation Refinery in Delaware City, Delaware, were killed by nitrogen asphyxiation while working to install an outlet piping ell at the top of a hydrocracker reactor. The hydrocracker unit was on turnaround, and the catalyst had been replaced in the reactor. As customary, a nitrogen flow purged the reactor with the nitrogen exiting the vessel from the open top flange. The large outlet piping ell had been removed to facilitate access to the reactor and accommodate the catalyst replacement. Due to the pyrophoric nature of the catalyst, nitrogen was required to prevent it from reacting with air and self-igniting. Please refer to Figure 11.3 for an image of the sign placed at the reactor opening. This sign is difficult to read but says, 'Confined Space – Do Not Enter'. There is no warning about nitrogen or nitrogen purge. A close-up image of the sign that was placed at the entrance to the vessel after the incident occurred is available in Figure 11.4.

A sign on the reactor indicated 'confined space', and red barricade tape had been placed around the bolts on the outlet flange. However, even though nitrogen was flowing through the catalyst and exiting from the large top flange, there were no nitrogen warning signs or barricades around the top of the reactor and nothing to alert the workers of the presence of nitrogen.

At 9:00 p.m., Valero issued a work permit to the Matrix contractors to 'Install the Top Elbow' on the reactor. The work permit specified 'Nitrogen Purge or Inerted' as NA, and 'Lockout/Tagout' was also marked as NA. There was some confusion around the work permit. The initial permit was for 'set up only', and there was no intent for the workers to enter the reactor at any point in the process. The workers proceeded to the top of the reactor and began preparations for installing the top elbow.

Figure 11.3 The Delaware City Refinery hydrocracker reactor top flange. *Source:* US Chemical Safety and Hazard Investigation Board/Public domain.

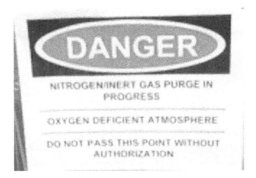

Figure 11.4 Close-up image of the safety sign placed at the reactor vessel's entrance after the incident occurred. *Source:* US Chemical Safety and Hazard Investigation Board/Public domain.

While waiting for the crane to lift the 5-ton elbow into position, one of the workers started cleaning the gasket surface on the top flange. He positioned himself by sitting on the top outlet flange with his legs dangling inside the reactor. He noticed a roll of duct tape lying on the distributor plate approximately 5 feet into the reactor. This would cause the reactor to fail the cleanliness inspection, so he located a piece of wire and formed a hook to remove the tape from the reactor. While 'fishing' for the duct tape, he was either overcome by the nitrogen and fell into the reactor or decided to enter the reactor to recover the duct tape. More likely, since he was directly in the breathing zone for the nitrogen, he was overcome and slid into the reactor. This is where the incident became even more tragic.

An eyewitness working on an adjacent vessel saw the job foreman peering into the reactor, then grabbing a nearby ladder and quickly entering the reactor, attempting to rescue his partner who was down in the reactor. Both matrix contractors were killed in this tragic incident. Figures 11.5 and 11.6 illustrate the roll of duct tape that the employee was attempting to retrieve and the wire being used.

This is not at all an uncommon story. Unfortunately, with nitrogen and toxins like H_2S, it is common to have more than one victim in these tragic scenarios. It is a very human trait that when

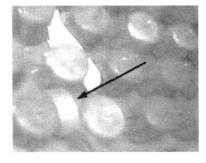

Figure 11.5 Roll of duct tape lying on the hydrocracker reactor distributor tray. This image clearly illustrates the roll of duct tape that was left on the distributor tray. This is what the craftsman was attempting to remove when he was overcome by the nitrogen atmosphere. *Source:* US Chemical Safety and Hazard Investigation Board/Public domain.

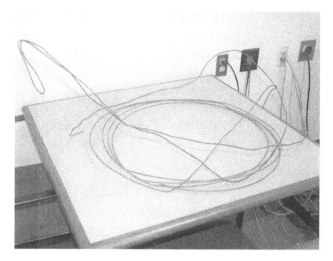

Figure 11.6 Roll of wire used by contractor in attempt to remove the duct tape from the reactor. *Source:* US Chemical Safety and Hazard Investigation Board/Public domain.

we see someone down, the urge is to attempt to rescue, and we unknowingly place ourselves in harm's way by entering the area without training and personnel protective equipment.

We just saw the same thing occur at the Foundation Food Group incident, where there were two fatalities in the incident, quickly followed by four other fatalities who were attempting to aid the first two victims and investigating the cause of the release.

The CSB quoted two different studies providing data on the fatalities of would-be rescuers as follows: 'One study reported that of 88 total confined space entry fatalities, 34 workers (39%) died while attempting to rescue a co-worker (*Journal of Safety Research* 1990). Another study reports that in eight incidents documented over 18 months, 10 rescuers died during rescue attempts. In two of the eight, the victim survived, but the rescuer died (NIOSH 1986)'.

Key Lessons Learnt

- **Work permits:**

    The work permit should be issued specifically for the task we are going to complete, and the hazards of the task should be communicated to the workers. In this case, we missed an opportunity to communicate the nitrogen hazard to the workers. Nitrogen purge was marked as 'not applicable' even though nitrogen was flowing through the catalyst and venting from the top flange directly where the contractors would be working.

    The CSB discovered that the work permit for this task was issued without performing the required safety-critical job site visit. It would be expected that a discussion of the nitrogen purge would have occurred during the job site visit had it occurred.

- **Employee training:**

    Employees should be trained in the hazards of the materials they will be working with and around. In this case, the workers were not made aware that the reactor was under a nitrogen purge.

- **Warnings and barricades:**

  Warnings and barricades should be positioned to alert the workers about the hazards associated with the job to be done. A sign warned the workers about a confined space, but there was no warning about nitrogen purge and the venting of nitrogen from the top manway.

- **Drills and exercises:**

  Workers should participate in periodic drills and exercises in realistic scenarios that they may experience while on the job.

- **Rescuing the victims:**

  When someone is observed to be down in the workplace where toxics or asphyxiants are used or processed, we should always call for help and avoid rushing in to provide assistance unless we are trained as responders and are properly equipped with the proper personnel protective equipment.

  It is a natural human response to want to rescue and help a teammate who has collapsed while working in a hazardous environment. This must be resisted, allowing our training to take over and an immediate call for help should be made. We should then stand out of harm's way and direct the responders to the site when they arrive.

11.3 What Has Happened / What Can Happen Incident Case Study Number 3 – Argon Asphyxiation

Alaska Welders Helper Asphyxiated in Argon-Inerted Pipe – One Fatality
Argon Asphyxiation
Incident Date: 29 April 1994

The US National Institute for Occupational Safety and Health (NIOSH) investigated this incident and issued a detailed report. Below is a summary of this tragic incident. See the Additional References section below for a link to the entire report. Unfortunately, although well-written, due to ongoing road construction in the area at the time, the report does not include images of the worksite or the accident scene.

At this Alaskan facility, a work crew consisting of a pipefitter foreman who was in charge of the work, a welding foreman, two welders, two operators, a labourer and three welders helpers was preparing to weld an interconnecting pipe to another section of the pipe to make up a gathering centre or flow station that was traversing under a roadbed. They were preparing for the weld by attempting to create an inert environment inside the piping using an isolation dam or internal pig designed to keep air (oxygen) away from the weld and a continuous purge of argon into the pipe as the inert gas. This is a regular and routine aspect of welding. The inert atmosphere prevents the oxygen in the environment from interacting with the weld metal and creating the formation of oxides and nitrites, which interfere with the weld quality and may cause the welds to fail the quality inspection.

Witnesses reported that the welders had been entering the pipe on a regular basis to back weld the pipe joints. Ventilation was in place at the opposite end of the pipe, about 70 feet away.

The workers were monitoring the oxygen concentration on the backside of the dam using an oxygen analyser to ensure the oxygen concentration was low enough for the welding to continue. During this process, the oxygen analyser indicated that the O_2 levels were too high, and the workers felt that the tubing to the analyser was too long. There were too many joints or splices, and air possibly entered the tubing, resulting in the high reading. A decision was made to replace the tubing.

While preparing to replace the tubing, the workers noticed that a welders helper's hard hat was near the opening to the section of pipe being welded, and they began looking for him. He was quickly discovered inside the inerted and downward-sloping section of the pipe. Quickly starting ventilation, the three workers entered the pipe, forming a human chain by interlocking their hands and feet. One of the rescuers was able to attach a rope to the victim, and they pulled the worker from the pipe section.

After getting the victim out of the pipe, the other workers noticed that the victim's face was purple and that he had no pulse. The workers started CPR immediately and made a radio call for additional help. They continued CPR until the emergency medical technicians (EMTs) arrived about 20 minutes later. However, it was apparent that the EMTs were not well trained and could not effectively use all the equipment they brought to the scene. CPR was continued, by either the worker or the EMTs, until the injured worker was transported to a local clinic, where the victim was declared dead. The victim's death was determined to be 'asphyxia by environmental suffocation'.

The NIOSH FACE investigation team concluded that the victim came into contact with the argon as he broke the plane of the pipe at the intersection of the horizontal segment and the beginning of the 45° downward slope. His position indicated that his unconsciousness may have occurred immediately after breathing the argon that had seeped through the dam.

A quick review of argon and its characteristics:
- Argon is a colourless, odourless and non-combustible gas. The victim may not even recognize that they are being overcome.
- It is heavier than air and can asphyxiate by displacement of air.
- Argon is an asphyxiant. Effects are due to lack of oxygen – lack of oxygen can kill.
- Moderate concentrations may cause headache, drowsiness, dizziness, excitation, excess salivation, vomiting and unconsciousness.

NIOSH FACE Investigation Recommendations. Note: The author felt these recommendations were right on the mark and included the full text here:

- **Employers and contractors should ensure that all permit-required confined spaces are identified and that an appropriate system for entry / work is in place.**
  Author comment: It is noteworthy that, although workers were routinely entering the confined space, the pipe was not considered a permit-required confined space, and procedures and protocols were not in place.
  Discussion: This work site was not considered to be a permit-required confined space. However, the investigation revealed that workers routinely entered the pipe to do back-welding operations. Although some controversy surrounds this issue, either the site was incorrectly evaluated as a confined space hazard, or the job was modified without the knowledge of safety and health personnel. Employees must understand what constitutes a confined space. Any modification to a job requiring confined space entry must be re-evaluated as a potential

hazard. When sites are identified as permit-required confined spaces, an appropriate protocol for entry and work must be instituted.

- **Employers and contractors should ensure all workers use appropriate confined space equipment and procedures. All workers entering a permit-required confined space must have an attached lifeline and a method for communicating with coworkers. Confined spaces must be properly ventilated prior to entry, and a 'competent person' must monitor the confined space entry and work operation.**

    Discussion: No permit-required confined space entry or work procedures were used in this incident. The site had been evaluated as non-permit required because the job was described as not requiring the entry of workers at any time. Standard confined space entry and work procedures may have prevented this fatality from occurring. The use of appropriate atmospheric testing prior to entry would have revealed an oxygen-deficient atmosphere. The use of intrinsically safe radio transceivers would have allowed ongoing communications during the operation. Any problem experienced by a worker would be identified earlier through requests for assistance by the worker or lack of response by the worker.

    General communications appear to have been flawed in some manner in this incident. The victim was apparently knowledgeable about the hazards of argon, yet he entered the inert space. He seems to have believed that the space had already been ventilated. Investigators suspect that the Pipefitter Foreman directed the victim to enter the pipe. However, the employee denied this assertion. Given that there are no direct witnesses to the conversation between the foreman and the victim prior to the incident, it is currently unclear why the victim believed the pipe to be safe.

    The use of a 'competent person' in a confined entry procedure could have prevented this fatality. A 'competent person' can concentrate only on the confined space task. Thus, divided attention difficulties are reduced. Also, a 'competent person' can quickly respond to problems encountered during the operation, such as activating a winch to rescue an injured worker. The use of a lifeline would have permitted safe retrieval of the victim, which would have been faster and not required unorthodox retrieval methods.

- **Employers and contractors should ensure that workers entering confined spaces know the appropriate procedures to rescue an injured worker in a confined space.**

    Discussion: Entry prior to atmospheric testing is extremely hazardous. The use of guesswork to estimate appropriate ventilation times is a dangerous procedure. Also, the use of a 'human chain' is not recommended. All of these problems could have been avoided through the use of appropriate confined space entry, work and rescue as described above. The procedures used in this incident could have resulted in the deaths of three 'would-be rescuers'. Adherence to standard methods results in quicker and safer retrieval of injured workers in confined spaces. The use of a retrieval harness attached to a (prior to routine entry) winch would have resulted in a fast, safe retrieval and would not have required the entrance of additional personnel into a hazardous situation.

- **Employers and contractors should ensure that emergency medical technicians are fully trained and competent prior to assigning such employees to regular duties.**

    Discussion: Witnesses described the EMTs called to the scene as confused. They did not appear to want to assist the victim's coworkers in performing CPR. They were unfamiliar with the operation of the defibrillator in the ambulance. One EMT was reading the defibrillator manual while CPR was being conducted. Investigation revealed that the two EMTs had just reported to work on the day of the fatality. Although responding to an industrial injury of this magnitude is a challenge for workers on the first day of a job, all EMTs must be fully trained and experienced prior to their assignment of sole responsibilities for emergency medical procedures. An alternative may be to assign new EMTs with experienced EMTs. It may be risky to assign such responsibilities to two new EMTs in an isolated environment, where access to a physician is extremely limited.

11.4 What Has Happened / What Can Happen Incident Case Study Number 4 – Argon Asphyxiation

Singapore Worksite Incident – One Fatality

Argon Asphyxiation

Incident Date: 24 November 2020

Background:

A worksite incident occurred in Singapore involving the asphyxiation of a worker by argon. This incident was investigated by the Ministry of Manpower in Singapore, and a Safety Bulletin was issued by the Workplace Safety & Health Council (WSH), Singapore. The safety bulletin was issued to share the incident with others as a way to help prevent a reoccurrence of the event. Thanks to the WSH for developing this important bulletin and allowing it to be used in this format. This was a well-written bulletin with good precautions when working with argon. It is carried forward essentially in its original form, albeit slightly modified to include some US OSHA regulatory requirements and guidance vis-à-vis entry into confined spaces. These additions are highlighted below as 'added text'.

In the United States, OSHA has additional regulatory requirements spelt out in the regulation that must be met: OSHA 29 CFR (1910.146). For these additional details, please refer to the regulation.

The incident:

On 24 November 2020, a worker was tasked with arc welding on a pipe. An inert gas purge was introduced into the pipe to facilitate the welding. The worker was subsequently found unconscious with his upper body inside the opening of an adjoining pipe. The worker was conveyed to the hospital, where he was pronounced dead. See Figure 11.7, which indicates where the welding was to be done and where the inert gas was to be applied. It also shows where the deceased worker's body was found.

The WSH Safety Bulletin states that the information contained is preliminary and subject to change. Lessons learnt are meant to enhance workplace safety and are not restricted to prevent sharing current lessons learnt. Please see the statement below:

> *Information on the accident is based on preliminary investigations by the Ministry of Manpower as of 21 December 2020. This may be subject to change as investigations are still ongoing. Please note that the recommendations provided here are not exhaustive, and they are meant to enhance workplace safety and health so that a recurrence may be prevented. The information and recommendations provided are not to be construed as implying any liability on any party, nor should it be taken to encapsulate all the responsibilities and obligations under the law.*

156 | *11 Nitrogen Asphyxiation Case Studies, and Asphyxiation by Other Inert Gases*

Figure 11.7 An image of the piping welding site where a welder was asphyxiated by argon while preparing to weld the adjacent piping and the location where the deceased worker's body was found. *Source:* From the Workplace Safety and Health Council.

On 24 November 2020, a worker was tasked to carry out arc welding on a pipe. To facilitate welding works, an inert gas was earlier introduced into the pipe. The worker was subsequently found unconscious with his upper body inside the opening of an adjoining pipe. The worker was conveyed to the hospital, where he was pronounced dead.

The WSH recommended that stakeholders such as occupiers, employers and principals in control of similar workplaces and work activities are advised to consider the following risk control measures to prevent similar accidents.

Ensure the safe use of inert gas for welding works:
The use of inert gases, particularly in confined spaces (likely the adjoining pipe in this case), presents an asphyxiation hazard as inert gases can dilute or displace oxygen from the air. When carrying out work activities that require handling inert gases, risk assessment (RA) must identify the risk of asphyxiation and the necessary controls that must be in place prior to starting work.

For gas-shielded welding work, the containment of inert gas (e.g., argon) in the weld area enhances the weld quality. This technique of inert gas containment is commonly referred to as purge damming. One way to carry out purge damming safely is to use commercially available inflatable purge dams or bladders. These dams or bladders are designed with pull cords that can help reduce or eliminate the need for workers to enter a pipe section.

Establish and implement safe work procedures (SWPs) for the selected method of purge damming and ensure that the SWPs are communicated to and understood by all relevant workers.

When inert gases are used, equip workers with a portable gas detector to detect the lack of oxygen and / or presence of toxics / flammables depending on the work environment.

The safety bulletin also recommended enhanced worker training and communication:
There are different types of welding (e.g., gas metal arc welding, gas tungsten arc welding), and each type may require a different work method. Workers must be trained for the specific type of welding they are tasked to perform.

Deploy only properly trained and qualified welders to carry out specialized welding works requiring the application of inert gases. Such welders must be aware of the hazards associated with inert gas handling and the risk controls to be taken. Instruct workers to never insert their heads into a confined space as long as the atmosphere is uncertain.

Excerpts from the WSH Safety Bulletin 'Ensure Safe Entry into Confined Spaces' with Added Text from the Respective OSHA Confined Space Standard

All confined spaces in a workplace must be clearly identified and labelled on-site.

According to the WSH (Confined Spaces) Regulations, a confined space can be any pipe in which the supply of air is inadequate or is likely to be reduced to be inadequate for sustaining life.

Added text: US OSHA defines a non-permit-required confined space as one that does not contain or, with respect to atmospheric hazards, have the potential to contain any hazard capable of causing death or serious physical harm.

Added text: US OSHA defines a permit-required space as one that has one or more of the following characteristics:

- Contains or has the potential to contain a hazardous atmosphere.
- Contains a material that has the potential for engulfing an entrant.
- Has an internal configuration such that an entrant could be trapped or asphyxiated by inwardly converging walls or by a floor that slopes downward and tapers to a smaller cross-section.
- Contains any other recognized serious safety or health hazard.

The WSH (Confined Spaces) Regulations define 'entry' as ingress by a person into a confined space, which occurs when the person's head passes through an opening into the confined space.

Added text: OSHA defines entry to mean the action by which a person passes through an opening into a permit-required confined space. Entry includes ensuing work activities in that space and is considered to have occurred as soon as any part of the entrant's body breaks the plane of an opening into the space.

Avoid entry into confined spaces and explore alternative, safer methods for carrying out work wherever possible. Where entry into a confined space cannot be avoided, the requirements stated in the WSH (Confined Spaces) Regulations must be adhered to.

Prior to confined space entry, carry out purging (to rid the atmosphere of hazardous contaminants) and ventilation (to render the environment safe for work). The atmosphere of the confined space must also be tested using suitable and calibrated instrument(s) for the level of oxygen content, level of flammable gas or vapour, and concentration of toxic gas or vapour, where applicable.

Entry into or work in a confined space may be allowed only if the purpose of entry and the atmospheric test results have been evaluated, and a confined space entry permit is issued by the occupier. Refer to the WSH (Confined Spaces) Regulations for the special conditions under which the confined space entry permit requirement does not apply.

Added text: The US OSHA regulation requires 'the employer shall verify that the space is safe for entry and that the pre-entry measures required by paragraph (c)(5)(ii) of this section have been taken, through a written certification that contains the date, the location of the space, and the signature of the person providing the certification. The certification shall be made before entry and shall be made available to each employee entering the space or to that employee's authorized representative'.

Work supervision and monitoring:

Provide on-site supervision to ensure that all SWPs with regard to the use of inert gases are strictly adhered to.

The whereabouts of workers should be monitored, especially those who have to work alone. Tracking methods may include signing in at key work locations or checkpoints, regular check-in through radio communications, monitoring through closed-circuit television, use of personal wearable devices and use of global positioning system to indicate worker location. Monitoring will allow employers to respond quickly should an emergency arise.

Emergency response plan:
Establish an emergency response plan for rescuing persons from the work area / confined space in the event of an emergency.

Ensure an adequate supply of rescue equipment such as breathing apparatus, safety harness, ropes and reviving apparatus.

Ensure that appointed rescue personnel have received adequate training, which can include first aid procedures and the use of rescue equipment.

Risk assessment:
Conduct a thorough RA for all work activities to manage any foreseeable risk that may arise when workers are required to work with inert gases. The RA should cover, but not limited to, the following areas:

- Presence of asphyxiation hazard.
  - Identification and control of entry into confined spaces.
  - Adequacy of on-site supervision.
  - Worker deployed to work alone (lone worker).
  - Emergency response / rescue plan.

Additional Resources (in Singapore)

- Workplace Safety and Health Act.
- Workplace Safety and Health (Risk Management) Regulations.
- Workplace Safety and Health (General Provisions) Regulations.
- Workplace Safety and Health (Shipbuilding and Ship-Repairing) Regulations 2008.
- Workplace Safety and Health (Confined Spaces) Regulations.
- Code of Practice on Workplace Safety and Health Risk Management.
- SS 510: 2017 Code of Practice for Safety in Welding, Cutting and Other Operations involving the Use of Heat.
- SS 568: 2011 Code of Practice for Confined Spaces.
- WSH Council's Technical Advisory on Working Safely in Confined Spaces.
- WSH Council's Workplace Safety and Health Manual for Marine Industries.

Additional International Resources

- UK HSE's Information Sheet: Asphyxiation Hazards in Welding and Allied Processes.
- UK HSE's Information Sheet: The Risks Posed by Exposure to Inerting Gases in the Open Air.
- International Maritime Organization's (IMO's) Revised Recommendations for Entering Enclosed Spaces Aboard Ships.

Additional Resources (in the United States)

- US Occupational Safety and Health Administration (OSHA) Code of Federal Regulations 29 CFR (1910.146) – "Permit Required Confined Spaces".
- US Occupational Safety and Health Administration (OSHA) Code of Federal Regulations 29 CFR (1926) (Subpart AA) – "Confined Spaces in Construction".
- US National Safety Council (NSC) "Confined Space Training for construction"
- OSHA's Factsheet on Controlling Hazardous Fume and Gases During Welding.
- https://www.nsc.org/getmedia/93676024-217d-4715-aa0f-03cd0e32bc7d/compliance-con fined-space-entry020322.pdf
- Airgas Safety Data Sheet (SDS) for Argon. https://www.airgas.com/msds/001004.pdf
- XO Safety – Example Confined Space Entry Checklist "Confined Space Entry Permit". https://view.officeapps.live.com/op/view.aspx?src=https%3A%2F%2Fcdn.shopify.com%2Fs%2Ffiles%2F1%2F0767%2F3985%2Ffiles%2FConfined_Space_Permit.docx%3F42769435038701780&wdOrigin=BROWSELINK
- OSHA.com "Online OSHA Training – Confined Space Entry – 8 Hour General Industry". https://www.osha.com/courses/confined-space.html#:~:text=This%20course%20addresses%20how%20to,the%20OSHA%20Standard%2029%201910.146

11.5   What Has Happened / What Can Happen Incident Case Study Number 5 – Carbon Dioxide Asphyxiation

Fatal Carbon Dioxide Exposure at the Smoley Mountain Opry Music Venue (One Fatality)
Pigeon Forge, Tennessee
Incident Date: 3 March 2018

Note: This case study was pulled together from several media reports and available information from the Tennessee Occupational Safety and Health Administration (TOSHA).

The Tennessee Opry is a music venue and performing arts theatre in Pigeon Forge, Tennessee, featuring some of the top country musicians and performers. The theatre opened in 1997 and operated until the COVID-19 virus struck. The venue closed in 2020 and has not since reopened.

On 3 March 2018, an accident occurred, which took the life of one of the theatre employees while he was preparing for one of the stage shows. At the Opry, carbon dioxide was used to generate the theatrical atmosphere and fog as a backdrop for the performance. Moona Rossol wrote a detailed article on the use of theatrical fog, its background and its effects titled 'Theatrical Fog, Smoke, and Haze Effects'. In that article, she lays out the chemicals used to generate the fog, and among those most commonly are carbon dioxide (dry ice) and liquid nitrogen. Other inert gases, such as argon, are also used in some cases. All of these are inert and, as discussed earlier, will displace the oxygen in the environment, especially when used in confined spaces or with minimal ventilation.

The Tennessee Occupational Safety and Health Administration (TOSHA) reported that the machine used to generate the fog effects for the theatre had been previously modified to allow additional fog to be produced. This modification included adding a manual ball valve and PVC pipe to allow carbon dioxide (CO_2) to be generated in the basement (an enclosed area) for a period of about 15–20 seconds before it would be needed for the performance. This change was made to ensure enough fog would be available when required for the performance.

On the day of the incident, it was reported that one of the managers was concerned about the timing and availability of the fog for the upcoming performance. It was also reported that the manager requested the employee to start the CO_2 at least a minute before it was needed to ensure enough fog would be available for a strong effect. The employee acknowledged that he understood.

During the performance, the employee missed his cue to start the fog, resulting in one of the other theatre employees going to the basement to check on his workmate. When he arrived, he found the employee having a seizure on the floor, apparently overcome by the carbon dioxide atmosphere in the basement. The responding employee radioed for help and then managed to get the CO_2 shutoff valve closed, stopping the flow of CO_2 into the basement. Another responding employee arrived, and they attempted to rescue the first employee. However, both responding employees were also overcome by the CO_2 gas. When EMS arrived, they found all three employees down and about 18% oxygen in the basement.

The theatre was immediately evacuated of the about 900 attending guests. The first employee died as a result of carbon dioxide asphyxiation (or actually due to a lack of oxygen). This is another example of 'would-be' rescuers attempting to rescue without proper training or equipment and becoming victims. As stated before, this is pretty common in cases of asphyxiation due to the lack of sensory warnings (no odour, no visual and no indication as to why the first victim is down).

The report summary is available on the US OSHA website link in the Additional Resources section. It is titled the 'Follow-up Federal Annual Monitoring Evaluation (FAME) Report'. The Tennessee summary report is in Appendix E.

11.6 What Has Happened / What Can Happen Incident Case Study Number 6 – Low Oxygen Content (Asphyxiation)

C&M Roustabout Services LLC

A Water Tank at a McClain County Well Site

Purcell, Oklahoma

Incident Date: September 2024

US Department of Labor investigation of worker's confined space fatality finds Oklahoma City-area contractor ignored federal safety measures, 2024/US Department of Labor/ Public Domain.

I included the following incident discussion from the OSHA and Department of Labor News Release, which was posted on March 18, 2024. This incident illustrates that incidents like those covered in this book are still occurring today.

OKLAHOMA CITY – An Oklahoma City employer's failure to follow federal safety procedures left a 30-year-old worker suffering fatal asphyxiation as they tried to make repairs inside a water tank at a McClain County well site in September 2023, a federal investigation has found.

Responding to the report of a fatality in Purcell, investigators with the US Department of Labor's Occupational Safety and Health Administration determined the worker had entered a permit-required confined space to fix a leaking bulkhead valve in a production tank and then lost consciousness. Two co-workers entered the tank in a failed rescue attempt and suffered effects from exposure to low atmospheric conditions. Neither sustained injuries.

OSHA investigators found the employer, C&M Roustabout Services LLC failed to evaluate the tank for hazardous conditions – including testing the atmosphere – and did not use protective systems to prevent worker injuries, a violation of federal regulations.

'This preventable tragedy must serve as a reminder of the importance of complying with safety and health standards', explained OSHA Area Director Steve Kirby in Oklahoma City. 'OSHA has specific requirements for operations such as drilling, servicing and storage to protect people employed in this high-hazard industry. Every employer should make safety and health a core value in their workplaces and ensure their workers are trained properly and provided the required safety equipment'.

Further investigation determined the remaining water in the tank contained compounds — including ethyl benzene, xylene, trimethylbenzene, isobutane and other cyclic aliphatic compounds — and atmospheric readings inside showed low oxygen content, volatile organic compounds and carbon dioxide, all of which can lead to asphyxiation.

OSHA issued C&M Roustabout Services 16 serious citations. Of these citations, 13 are related to failures that contributed to the worker's death, including the following:

- Allowing employees to enter a confined space without an atmospheric evaluation or a required permit.
- Not providing flame-resistant personal protective equipment.
- Failing to have testing and ventilation equipment space entry programs in place.
- Not training employees to establish proficiency on confined space hazards.
- Failing to ensure those authorized to enter the space knew the hazards, signs or symptoms of exposure.

11.7 What Has Happened / What Can Happen Incident Case Study Number 7 – Employee Dies due to Asphyxiation from Oxygen Displacement

Meriwether Compressor Station

1461 Meriwether Road, Browning, Montana

Incident Date: 10 May 2023 (Employee's body was discovered on 11 May 2023)

Energy Contract Services, LLC (One Fatality)

Inspection Information Search, 2024/US Department of Labor/Public Domain.

On the evening of 10 May 2023, an employee working for Energy Contract Services, LLC, at the Meriwether Compressor Station on 1461 Meriwether Road in Browing, Montana, was killed.

(a)

(b)

Figure 11.8 a and b The Meriwether Compressor Station, serving northern Montana. *Source:* NorthWestern Energy/https://www.facebook.com/NorthWesternEnergy/photos/the-meriwether-compressor-station-in-northern-montana-is-a-great-example-of-how-/2314534555254265/ Last accessed on November 20, 2024.

The Meriwether Compressor Station is a relatively small compressor station for recovered natural gas from nearby wells and is operated by NorthWestern Energy.

The employee had been pressure testing a section of piping with nitrogen and was apparently overcome by the inert gas. His co-workers, who had left him the previous evening, found him the following morning. The employee told his co-workers that he would finish the job and that it was okay for them to leave. Figure 11.8a and b illustrate the buildings and location where this incident occurred. Please note that, especially in cold weather, there appears to be little air circulation in these buildings.

What is known about this incident is that the pipe was found open with nitrogen leaking into the surrounding area. US OSHA investigated the fatality and reported that the employee was asphyxiated due to the displacement of oxygen in the surrounding environment with nitrogen, resulting in the employee's death.

What can be learned from this tragic incident:

- The employee was working alone with nitrogen, a known asphyxiant. In a case like this, it is best to have a co-worker who can help with safety oversight. If the co-worker is properly trained, they can also provide assistance or they can alert trained rescuers if necessary, in the event someone is overcome.

- The question of equipment isolation and lockout and tagout also comes into play. It is obvious from the OSHA report that nitrogen was being released from the open pipe into the surrounding area. This raises a couple of important questions. Was the piping properly isolated and depressurized? Why was nitrogen leaking into the surrounding area?
- Also, was a procedure available for this task and was it being followed?

11.8 What Has Happened / What Can Happen Incident Case Study Number 8 – An Explosion Occurred While Unloading Liquid Nitrogen at an Ice Cream Facility (10 Injured)

Lion Technology posted a report by Lauren Scott and Roseanne Bottone on 24 July 2021, about an incident that occurred on 21 July 2021, while unloading liquid nitrogen at an ice cream facility in Kentucky. This report included several factors from the above that all came together, resulting in an explosion and 10 people being hospitalized due to their injuries. There is a link to this report in the Additional Resources section.

The report indicated that the 10 employees were recovering from their injuries and that the cause of the explosion was still under investigation. The site where the unloading incident occurred was owned by a nationwide ice cream chain; the site typically makes ingredients for a third-party company.

Ten employees are recovering after sustaining injuries related to a liquid nitrogen explosion at a Kentucky ice cream facility last week, on 21 July. The facility used liquid nitrogen to flash-freeze the ice cream products. The incident occurred when a truck was unloading liquid nitrogen into a storage tank at the facility. At the time of the report, the cause of the explosion was under investigation.

Nine workers were released from local hospitals the day after the explosion, with one still undergoing medical care. Although the facility is owned by a nationwide ice cream chain, the site where the incident occurred typically makes ingredients for a third-party company. The facility uses liquid nitrogen to flash-freeze ice cream products.

The report confirms that when liquid nitrogen vapourizes, it expands by a factor of nearly 700. For example, 1 litre of liquid nitrogen expands to fill nearly 25 cubic feet when it evaporates to form nitrogen gas, which can easily result in an explosion like the one that occurred in Kentucky. The report also confirms that a release of nitrogen will displace the oxygen and can cause suffocation.

11.9 What Has Happened / What Can Happen Incident Case Study Number 9 – Fatality of Welder in Confined Space Welding in the Presence of Argon

This incident was investigated by the Mine Safety Administration, US Department of Labor. A final report was issued on 15 September 2021. The following is a summary of this report. For additional details and a copy of the final report, please see the link in the Additional Resources section.

This accident happened at the Freeport–McMoRan Morenci Inc. copper mine located in Greenlee County, Arizona, when a 33-year-old welder with over 10 years of experience entered a 30-inch-diameter section of piping to check a weld. Argon was being used as an inert gas in a dammed-off area inside the piping to shield the weld from oxygen in the environment, protecting the weld from oxidation and corrosion. Unfortunately, the argon resulted in the welder's asphyxiation.

Two welders had been welding inside the pipe and noticed some weld deterioration and suspected a less-than-adequate argon shield. One of the welders replaced the argon cylinder with a full one and resumed the argon purge into the dammed-off area. The welder missed his workmate and, with help, located him unresponsive, approximately 40 feet into the 30-inch pipe. With additional help, the unresponsive welder was pulled from the pipe, where first aid, including resuscitation, was administered. Remaining unresponsive, he was transported to the hospital, where he was pronounced dead.

The investigation determined that the contractor did not have an effective confined space entry procedure or policy. The welder entered the confined space without being trained in the hazards and safety precautions associated with working in a confined space with argon gas. The atmosphere inside the 30-inch pipe had not been tested to determine a safe atmosphere for entry, and the welder was not equipped with breathing air protection. The welder also did not have a safety belt or rescue lines, nor was there a backup person monitoring the welder during the entry. No warning signs or barricades were present to warn other miners from entering confined spaces.

In a confined space with limited air circulation and without breathing air, argon works on the human body exactly as nitrogen. It is an inert gas and quickly displaces the oxygen resulting in either hypoxia or asphyxiation and can lead to death in a matter of minutes. The hazard is that the victim is totally unaware that this is happening. They simply go to sleep and sometimes never wake up. Unfortunately, this is exactly what happened at this mine when the welder ventured into the 30-inch pipe purged with argon and was overcome, not by the argon, but by the lack of oxygen.

The accident investigation team identified the following underlying causes of the accident:

- The contractor did not train the contract miners in the hazards and proper safety precautions concerning argon gas.
- The contractor did not verify a respirable atmosphere before a contract miner entered a confined space (no gas test to confirm oxygen level before and during entry).
- The contractor did not confirm that contract miners entering confined spaces were wearing safety belts and lines and have an additional miner to monitor and adjust their lines, as necessary (no rescue plan for confined space entrants).
- The contractor did not erect barricades or warning signs to prevent or warn miners from entering confined spaces.

The contractor company was cited by the US Mine Safety and Health Administration for each of the underlying causes of the accident. The contractor reportedly has taken corrective actions to address each of the underlying causes.

11.10 What Has Happened / What Can Happen Incident Case Study Number 10 – A Summary of Incidents Involving 'Would-be Rescuers'

Background:
We have already documented in this book several incidents where other personnel or workers witnessed someone enter a vessel with an inert atmosphere or collapse into a vessel. They either

rescued or attempted to rescue the individual(s). Unfortunately, this typically results in the would-be rescuers becoming victims by succumbing to the inert atmosphere. Unfortunately, this happens all too often. This is a natural human instinctive reaction to try and help a workmate or someone else in trouble. Unfortunately, this results in additional people who need to be rescued and may delay response for the initial victim.

This chapter includes a list of the known cases where this has occurred. Of course, this is not intended to be a complete list, and I believe there are many other examples that we have yet to hear about.

We have stressed before and will do so again here. One should never attempt to enter a vessel without proper training and proper rescue equipment. All procedures for confined space entry must be in place, including a test of the atmosphere in the vessel or confined space, a standby person in communication with those inside the vessel, and all of the additional confined space entry procedures as outlined in Chapter 4.

There are only several cases of workers being killed while attempting to rescue others highlighted here; however, many others similar to these are included in the Appendix.

- **Fatal Liquid Nitrogen Release at Foundation Food Group (Six Fatalities, Four Serious Injuries)**
  Gainesville, Georgia
  Incident Date: 28 January 2021
  In this case, two workers were killed in the initial nitrogen release. According to the Chemical Safety Board, 14 others, including members of management, entered or came close to the confined space. Six of them were also killed.
  The incident claimed a total of six fatalities, including four would-be rescuers.
- **Fatal Nitrogen Exposure at Valero Delaware City Refinery (Two Fatalities)**
  Delaware City, Delaware
  Incident Date: 5 November 2005
  In this case, the Chemical Safety Board reported that one of the workers either entered a reactor vessel under nitrogen purge to remove a roll of duct tape or was working near the open manway and was overcome and fell inside.
  We know that the second person, a workmate, realized the first victim was inside the reactor and was motionless and purposely entered the reactor with the intention of rescuing. Unfortunately, he also died due to asphyxiation.
  There were two fatalities in this incident.
- **Alaska Welders Helper Asphyxiated in** Argon-Inerted **Pipe (One Fatality)**
  Argon Asphyxiation
  Incident Date: 29 April 1994
  In this incident, the workers attempted to weld inside a downward-facing pipe, positioned dams near the welds, and purged the pipe with argon to facilitate the welding. Oxygen was being measured with an O_2 detector, which showed that the O_2 level was too high. They concluded that the analyser sample tubing was leaking at the joints and devised a plan to replace it. Unfortunately, a short while later, a welder helper was discovered motionless and non-responsive inside the pipe. The pipe was quickly purged to remove the argon, and three other workers entered the downward sloping and pulled the victim out. CPR was administered, and he was transported to the hospital, where he was declared dead.
  This incident resulted in one fatality, but three would-be rescuers put themselves at risk.

- **Fatal Carbon Dioxide Exposure at the Smoley Mountain Opry Music Venue (One Fatality)**
  Pigeon Forge, Tennessee
  Incident Date: 3 March 2018
  A theatre worker purged what appeared to be an excessive amount of liquid nitrogen into a backstage room in support of a theatrical production and was overcome by a lack of oxygen. Two other workers entered the nitrogen-filled room, turned off the device that was releasing the nitrogen, and were overcome by a lack of oxygen and required medical intervention.
- **December 2020 US OSHA Report**
  California Ranch Food Company, Inc. (Two Fatalities)
  At 7:45 p.m. on 1 December 2020, Employee no. 1, a production lead supervisor, entered the chill / production room. She observed a cloud generating in the immediate area of the freezer tunnel. She went to investigate the cause of the cloud and was killed by asphyxiation. Approximately one hour later, Employee no. 2, a production supervisor, entered the chill / production room. He observed Employee no. 1 on the floor and went over to her. Employee no. 2 was overcome by the lack of oxygen in the room and was also killed by asphyxiation. Liquid nitrogen released from the tunnel freezer, Linde Cryowave Freezing tunnel Model 1250-5, in the chill / production room created an oxygen-deficient atmosphere that asphyxiated both employees.
- **February 2020 US OSHA Report**
  Kenan Advantage Group, Inc. / Great Lakes Tank & Vessel LLC (Two Fatalities)
  At 12:30 p.m. on 20 February 2020, Employee no. 1, employed by a structural steel fabricator and erector company, was entering a tank to clean it. The tank had a combination of Ecocure II and methyl ethyl ketone residues and had been purged with nitrogen. Employee no. 1 entered the permit-required confined space that contained the residual chemicals and nitrogen to perform the cleaning operations. She was overcome by the oxygen-deficient atmosphere. Employee no. 2, employed by a chemical distribution company, entered the tank to make a rescue attempt for Employee no. 1. He was also overcome by the oxygen-deficient atmosphere. Both employees were killed by asphyxiation.
- **December 2015 – US OSHA Report – SPX Transformer Solutions, Inc. (Two Fatalities, One Hospitalized)**
  On 30 November 2015, Employee no. 1 entered a large transformer, top cover removed, which had previously been filled with nitrogen. Employee no. 1 fell unconscious. Employee no. 2 entered to retrieve Employee no. 1 and also fell unconscious. Employee no. 3 followed and also fell unconscious. Employees nos. 1 and 2 died. Employee no. 3 was hospitalized.

The key lessons learned in this section are to ensure that first and foremost, the tank, vessel, or other enclosure is closely examined to see if it meets the OSHA definition of a confined space and especially a permit-required confined space, as outlined in the OSHA regulation (see Chapter 4).

- The OSHA regulation provides two definitions of a confined space as follows:
  - Non-permit confined space means a confined space that does not contain or, with respect to atmospheric hazards, have the potential to contain any hazard capable of causing death or serious physical harm.
  - Permit-required confined space (permit space) means a confined space that has one or more of the following characteristics:

1. Contains or has the potential to contain a hazardous atmosphere.
2. Contains a material that has the potential for engulfing an entrant.
3. Has an internal configuration such that an entrant could be trapped or asphyxiated by inwardly converging walls or by a floor that slopes downward and tapers to a smaller cross-section.
4. Contains any other recognized serious safety or health hazard.

If it is defined as a permit-required confined space, this means the following should be in place before entry occurs (not intended to be a complete list):

- If employees are to enter a permit-required confined space, the employer must develop and implement a written permit space program that fully complies with the regulation.
- The space shall be prepared by purging, inerting, flushing or ventilating as necessary to eliminate or control atmospheric hazards.
- Danger signs must be posted to inform employees and others of the potential hazards and to keep people at a safe distance from the openings. The regulation provides examples of the wording to be used.
- When openings or covers are removed, the openings shall be protected by guards, railing, a temporary cover or other barrier to prevent an accidental fall into the opening.
- A gas test should be conducted confirming there is adequate oxygen (a minimum of 19.5%), that toxics are identified and are within the limits approved by the OSHA standards and the lower explosive limit is no more than 10%. The atmosphere within the space shall be periodically tested as necessary to ensure the ventilation prevents a hazardous atmosphere. If a hazardous atmosphere is detected, employees shall leave the space immediately.
- A pre-entry written certification shall be developed to include the date, the space location, and the signature of the person providing the certification before entry into the permit-required space.
- Provide at least one attendant outside the confined space and in communication with all those in the confined space. The attendant must be capable of calling for help or support in an emergency.
- The entry permit must be retained for at least one year to facilitate the review of the permit-required confined space programme.

Again, this is a comprehensive standard and details a full range of requirements that must be in place to be in compliance with the regulation. This is a summary and is not intended to be a complete list. Please refer to the regulation for detailed information. The regulation is available in Chapter 4 or online at this link: https://www.osha.gov/laws-regs/regulations/standardnumber/1910/1910.146

End of Chapter 11 Review Quiz

1   What is by far the most common asphyxiant in general industry, the one that results in more deaths than most of the others combined? Why is this?

    Answer(s):

2 Following nitrogen, what is most likely the second most frequent gas that is involved in employee or worker fatalities? Why?

Answer(s):

3 What was the cause of the tragic incident that occurred at the Foundation Foods poultry processing facility on 28 January 2021?

How many people died in this tragic accident?

Answer(s):

4 At the Valero Delaware City Refinery, what information did the workers have that informed them that the reactor was being purged with nitrogen? Please explain your answer.

Answer(s):

5 How could the incident at Foundation foods have been prevented, or at least how could the consequences and loss of life have been prevented?

Answer(s): Please list all that you can.

6 What is the most likely outcome if one worker enters a confined space and is overcome by the lack of oxygen and his workmate decides to enter the confined space to attempt a rescue (without training and without specialized rescue equipment)?

Answer(s):

7 As defined by the OSHA regulation for confined space entry, what must be in place before an entrant can enter a confined space if it meets the definition of a permit-required confined space?

Answer(s): Please list as many of the requirements as you can.

8 US OSHA has two definitions for confined space: a non-permit confined space and a permit-required confined space. What is the major difference between the two definitions?

Answer(s):

9 We know that nitrogen and argon have no odour or other way we can be made aware of their presence and that the oxygen has been displaced, making these extremely dangerous, especially in a confined space or where there is restricted or no ventilation.

Is this also true for carbon dioxide (CO_2 or dry ice)?

Answer(s):

Additional References

National Library of Medicine, National Center for Biotechnology Information Environmental Gas Displacement: Three Accidental Deaths in the Workplace. https://pubmed.ncbi.nlm.nih.gov/11953489/

Monona Rossol, Care of the Professional Voice, Theatrical Fog, Smoke, and Haze Effects. https://www.nats.org/_Library/JOS_On_Point/JOS_077_5_2021_645.pdf

Follow-up Federal Annual Monitoring Evaluation (FAME) Report, Tennessee Department of Labor and Workforce Development Division of Occupational Safety and Health FY 2018, Tennessee Annual Report, Appendix E- FY 2018 State OSHA Annual Report (SOAR). https://www.osha.gov/sites/default/files/2019-06/tennessee_2018.pdf

Workplace Safety & Health Council (WSH), Singapore, Accident Advisory: Worker Found with Upper Body in Pipe. February 22, 2021, Ref: 2021092. https://www.tal.sg/wshc/-/media/TAL/Wshc/Resources/Newsletters/WSH-bulletins/Files/WA20210222pipe.pdf

US Chemical Safety and Hazard Investigation Board, Final Incident Report – "Valero Delaware City Refinery Asphyxiation Incident".https://www.csb.gov/valero-delaware-city-refinery-asphyxiation-incident/

US Chemical Safety and Hazard Investigation Board, Incident Reports – Union Carbide Nitrogen Asphyxiation Incident. https://www.csb.gov/assets/1/20/csb_unioncarbidefinal.pdf; https://www.csb.gov/assets/1/20/final_union_carbide_report.pdf

Practical Welding Technology, Rudy Mohler. Industrial Press, Inc. 1983.

Third International Conference on Welding and Performance of Pipelines, Edited by P.H.M Hart. The Welding Institute, 1987.

Occupational Safety and Health Administration, Permit-Required Confined Spaces for General Industry January 14, 1993. Final Rule (29 CFR Parts 1910).

Mine Safety and Health Administration, September 15, 2021 Fatality – Final Report, Accident Report: Fatality Reference. https://www.msha.gov/data-reports/fatality-reports/2021/september-15-2021-fatality/final-report

US Chemical and Hazard Investigation Safety Board

Final Incident Report – 'Fatal Liquid Nitrogen Release at Foundation Food Group'

Gainesville, GA

Incident Date: 28 January 2021.

https://www.csb.gov/investigations/completed-investigations/?F_InvestigationId=3616

12

Summary of Additional Actions to Help Prevent Asphyxiation Incidents at Our Facilities

The following is a summary of actions we can take at our facilities to help prevent oxygen deficiency or fatalities. The Department of Energy published a series of follow-up actions after several oxygen deprivation incidents at their sites. I felt that those applied equally to our facilities, and I have included them here along with several of my own.

- Ensure your facility has a specialized plan and procedure for permit-required confined space entry that includes all the elements and requirements outlined in the Occupational Safety and Health Administration (OSHA) regulation 29 CFR (1910.146).
- Advise workers to never connect a personal respirator device to a facility pipeline, even if the pipeline is labelled as air.
- Always verify that the breathing air being used is certified as grade D breathing air and verify the oxygen content before its use.
- Always ensure that workers follow the process to ensure the atmosphere is safe before entering any process storage tank or vessel. This means that before entering a vessel, verify that all hydrocarbon sources are effectively isolated, all hazardous material has been removed, the atmosphere has been tested for hydrocarbons (lower explosive limit), toxins, and oxygen deficiency, and an entry permit has been issued.
- Ensure workers understand that unless they are trained responders and are equipped with proper rescue equipment, including breathing air, they should always call for help instead of attempting to rescue a downed co-worker.
- Ensure all workers understand that they are not to transport nitrogen or any other compressed gas cylinders inside any vehicle.
- Recognize that oxygen-deficient atmospheres can be dynamic and exist outside of confined spaces. Analyse all potential facility conditions and implement protections with the hierarchy of controls.
- Design and engineer facilities to prevent or control oxygen-deficient atmospheres using interlocks, ventilation, gas controls and robust monitoring and alarms. Warning signs are never enough.
- Ensure facilities are designed so workers can hear or see alarms and lights at a safe distance from hazardous conditions.
- Periodically test warning alarms to ensure they are fully functional and that workers can see or hear them when activated. Alarms should also be incorporated into worker training and drills.
- Conduct routine facility inspections to ensure engineered systems will function properly and as expected (e.g. ventilation, valves and alarms are set correctly and are functional).

- Ensure that all workers know that they should never attempt to peek inside a storage tank, a tank truck or a process vessel immediately after vessel opening. A valid gas test must be completed before this is allowed.
- Track and confirm the implementation and effectiveness of controls to address deficiencies and support continuous improvement.
- Stop work if unexpected conditions are encountered. Exit potentially hazardous areas until confirmed safe.
- Ensure all procedures and postings, including building emergency plans, are correct and consistent, informing workers of the actions to be taken when an alarm is activated.
- Train affected employees on response procedures, including awareness of each alarm's visual and/or audible signals and actions to take if an alarm is activated.
- Conduct emergency drills to reinforce response to abnormal/alarm conditions.

End of Chapter 12 Review Quiz

1. What is each facility required to have in place before an entry into permit-required confined space can be executed?

   Answer(s):

2. Please provide the primary reasons why policies or detailed procedures should be in place that prevent workers from connecting an airline-supplied respirator to a facility pipeline for use as breathing air.

   Answer(s):

3. Always ensure that workers follow the process to ensure the atmosphere is safe before _____ any process storage tank or vessel. This means that before entering a vessel verification that all hydrocarbon sources are effectively isolated, all hazardous material has been removed, the atmosphere has been tested for hydrocarbons (lower explosive limit), toxins and _____ deficiency, and an entry permit has been issued.

   Answer(s):

4. Why is it important for workers and others to understand that they should never transport nitrogen cylinders, or pressurized cylinders of any type, in the cab of a pickup truck or any other enclosed vehicle, even in the trunk of a vehicle?

   What makes this an extreme hazard for those in the vehicle with the pressurized cylinders?

   Answer(s):

5. How should the space be verified to be free of hazards before an entry into a permit-required confined space can be attempted?

   Answer(s):

6. How can workers be made aware of alarms and the actions to take in the event an alarm is activated?

   How can this be enforced on a routine basis?

   Answer(s):

7. What actions should be taken if an entry into a permit required space is ongoing and an alarm sounds, or other similar unexpected event or condition arises?

   Answer(s):

8. Is it possible for low oxygen conditions to exist outside of confined spaces, especially when nitrogen purge is underway?

   How should we respond when this is recognized?

   Answer(s):

9. What actions should workers take if they see an associate collapse while working inside a permit-required confined space?

   Should they attempt a rescue?

   Answer(s):

10. What are the important design parameters for emergency gas detection alarms?

    Answer(s): Answer all you can.

Additional References

Occupational Safety and Health Administration, OSHA Technical Manual (OTM) Section VIII: Chapter 2, Respiratory Protection. https://www.osha.gov/otm/section-8-ppe/chapter-2

Occupational Safety and Health Administration, Permit Required Confined Spaces, 29 CFR 1910.146.

Occupational Safety and Health Administration, Respiratory Protection Standard, 29 CFR.1910.134.

The US Department of Energy, Oxygen Deficient Atmosphere Hazards at DOE Facilities – September 2021. https://www.energy.gov/sites/default/files/2021-10/OE-3_2021-05_Oxygen_Deficient_Atmosphere_Hazards_at_DOE_Facilities.pdf

13

Additional Discussion of Liquid Nitrogen Use in Ice Cream Shops

I started this book in the Foreword, and in several chapters, mentioning the hazards of nitrogen in ice cream shops as a refrigerant to flash-freeze the ice cream. Of course, this is primarily due to some shops having a large liquid nitrogen storage tank inside the shop where adults and children of all ages are present at all times of the day and evening. One article I read said that one shop that was visited had not one tank but four large liquid nitrogen storage tanks. I fear a leak in even one liquid nitrogen storage tank in an enclosed space like an ice cream shop can expose people to nitrogen asphyxiation or other safety concerns.

There are other significant hazards to the workers in these shops as they are working with liquid nitrogen, usually without much, if any, training on what they should do in the event of a release or exposure to themselves or others, and generally without much in the form of personal protective equipment. I am also concerned with the properties of the nitrogen being colourless and odourless and typically at an extremely cold temperature, typically colder than −300 °F below zero (−320 °F or −195 °C).

There are also many cautions and warnings on the web about the possibility of ingesting even a small amount of liquid nitrogen while eating ice cream made with nitrogen. The issues are the supercold liquids, the rapid and violent expansion of liquid nitrogen as it evaporates and the damage it can do to the intestine. Liquid nitrogen expands about 700 times as it evaporates. There is probably enough said about this, but it is a reality, and the preparer and the consumer must verify that there is no liquid nitrogen left in the ice cream before it is eaten. The US Food and Drug Administration is warning the public about the dangers of eating liquid nitrogen-infused treats (see Additional Resources section for more details). Also, please see the Additional Resources section for public reports of injuries involving ingesting ice cream and/or other cold treats made with liquid nitrogen.

The following are excerpts from the 2017 FDA Food Code (please see the Additional Resources section for a link to this code and more information):

> "However, liquid nitrogen and dry ice must not be used in ways that make food unsafe for consumers or that cause other safety hazards. Safety concerns associated with the use of liquid nitrogen and dry ice in the preparation of food and beverages at retail are based on both the physical state of the substances and accidents surrounding their use rather than on any toxicity associated with either substance. Both liquid nitrogen and dry ice can cause severe damage to skin and internal organs if mishandled or accidently ingested due to the extremely low temperatures they can maintain. As such, liquid nitrogen and dry ice should not be directly consumed or allowed to directly contact exposed skin."

"While retail food-related incidents of accidental ingestion or direct contact with liquid nitrogen and dry ice in the United States have been low, injuries have been severe. On August 30, 2018, FDA issued an advisory warning consumers and retailers of the potential for serious injury from eating, drinking or handling food products prepared by adding liquid nitrogen immediately before consumption as the liquid nitrogen may not completely evaporate before reaching the consumer or may leave the product at an extremely low temperature, posing a significant risk of injury."

A detailed analysis of the history and injuries sustained by nitrogen use in food and beverages was documented in 'A Qualitative Risk Assessment of Liquid Nitrogen in Foods and Beverages'. This was led by a group of scientists from the Canadian Research Institute for Food Safety, Dept. of Food Science, the University of Guelph, Guelph, Ontario, Canada, the Public Health Ontario, the Dalla Lana School of Public Health and the University of Toronto. This study contains case studies and peer reviews for 17 incidents of ingestion of liquid nitrogen contained in food and beverages.

The case reviews document some mild outcomes to very serious and potentially life-threatening effects of the ingestion of liquid nitrogen. The study found that if liquid nitrogen is ingested prior to its evaporation, there is a significant risk of gastrointestinal barotrauma (tissue injury caused by a change in pressure, which can compress or expand gas that is contained in various body structures and perforations). The report found that in 10 of the 17 cases, this was the primary injury reported. The report also cautions that the findings should be interpreted carefully since perforations are more likely to be reported.

The study also found that consumers may face many different types of injuries from ingesting liquid nitrogen. Inhalation and ingestion tend to be the most severe and often require surgery. Barotrauma with gastric perforation is the most severe outcome they identified. The risk assessment contains a list of 43 additional references, which provide a wealth of additional information on these hazards.

These concerns and others have been discussed throughout this book. Using liquid nitrogen to freeze ice cream is a relatively new technology, and we don't have much incident experience to review. To date, there have been severe injuries and near asphyxiations, but fortunately, no fatalities, at least, none that I am aware of. The closest case is what happened at the Foundation Food Group, where six people died, four of whom were attempting rescue. While little real data are available, the following summarizes the industry injury history.

I found it interesting that I had previously searched the OSHA injury database for all nitrogen-related injuries or deaths, and none of the following were reported. Please see a listing of the OSHA nitrogen-related incidents in Appendix 3, most of which meet the OSHA injury/illness reporting guidelines. Of course, a couple of the following incidents did not result in an injury, and those affected were not employees; therefore, OSHA reporting was not required in those cases.

Generally, OSHA considers eating on the job to be a personal activity, and therefore injuries while eating at work are not considered reportable. However, if the food was provided by the employer or if it was contaminated on the jobsite, this would be a recordable injury.

Injuries or incidents related to liquid nitrogen use in ice cream shops:

- Ice Cream Shop – Key Largo, Florida
  Approximately January 2024

Piping broke on storage overnight, releasing liquid nitrogen into the shop.
No one was present, and no injuries occurred.
The lettering below the ice cream counter was erased by the release of the supercold liquids.

- Ice Cream Shop – Weston, Florida
  6 June 2019
  Ice Cream Shop in Weston, Florida
  Liquid nitrogen leak from a large storage tank.
  One worker collapsed unconscious, and two responders, a firefighter and a deputy sheriff were also overcome. One person was admitted to the hospital for treatment.
  The news video of this incident shows the glass shop windows completely iced over.

- Dippin Dots Ice Cream Shop – Kentucky
  21 July 2021
  A liquid nitrogen explosion occurred while unloading liquid nitrogen at an ice cream facility in Paducah, Kentucky.
  Ten people were hospitalized due to their injuries; nine were released the following day.

- Colorado Springs, Colorado, a New Shop in a Strip Mall
  26 August 2024
  A leak occurred in a new 55-gallon liquid nitrogen storage tank.
  Emergency responders called and mitigated the situation.
  No injuries.

End of Chapter 13 Review Quiz

1. We learned earlier that two of the main hazards of liquid nitrogen are the potential for cold burns upon contact and asphyxiation from oxygen displacement. What is the third serious hazard with liquid nitrogen?

   Answer(s):

2. Liquid nitrogen is typically supercold (_____ °F) and can cause catastrophic burns to the digestive tract. When it vaporizes, liquid nitrogen expands _____ times, resulting in extreme internal injuries.

   Answer(s):

3. One of the most tragic nitrogen incidents to occur in recent times happened at the _____ _____ _____, where six people died, four of whom were attempting rescue.

   Answer(s):

4. An injury or illness resulting from ingesting liquid nitrogen while eating nitrogenated ice cream on the job is not reportable to OSHA.

   Answer(s): Please mark the correct answer.

   True

   False

5 Workers who handle liquid nitrogen or work near equipment containing liquid nitrogen should be properly _____ and wear proper _____ _____ _____.

Answer(s):

6 A leak in a liquid nitrogen storage tank in an enclosed space like an ice cream shop can expose people to _____ _____ or other safety concerns.

Answer(s):

Additional Resources

Ali D., Farber J., Kim J., Parto N., and Copes R. (2021). A Qualitative Risk Assessment of Liquid Nitrogen in Foods and Beverages. *Food Protection Trends*. 41. 293. 10.4315/1541-9576-41.3.293.

US Food and Drug Administration, Reference Document: 2017 FDA Food Code, Provision: 3-202.12 Additives; 3-302.14 Protection from Unapproved Additives, Document Name: Liquid nitrogen and dry ice in food, Date: October 31, 2018. https://www.fda.gov/media/117281/download#:~:text=Notices%20should%20instruct%20consumers%20to,conditions%2C%20physical%20abnormalities%2C%20and%20possibly

CALOX, Is Liquid Nitrogen Food Safe? April 15, 2022. https://caloxinc.com/blog/is-liquid-nitrogen-food-safe/#:~:text=FDA%20Recommendations&text=Accidental%20contact%20exposure%20to%20liquid,to%20the%20extremely%20low%20temperatures

University of Arkansas, U.S of A Food Scientists Caution Against Using Liquid Nitrogen With Kids' Foods. https://news.uark.edu/articles/42454/u-of-a-food-scientists-caution-against-using-liquid-nitrogen-with-kids-foods

Weston, Florida, Ice Cream Shop Nitrogen Leak and Release, ABC7 News. https://www.mysuncoast.com/video/2019/06/07/nitrogen-leak-ice-cream-shop/

Weston, Florida, Ice Cream Shop Nitrogen Leak and Injuries, Weston, Florida, WPLG News 10 Florida. https://www.local10.com/news/2019/06/07/3-sickened-after-being-exposed-to-nitrogen-gas-at-weston-ice-cream-shop/

Injured in Nitrogen Explosion at Dippin' Dots Facility in western Kentucky, Louisville Courier Journal (Article by Ayana Archie). https://www.courier-journal.com/story/news/local/2021/07/21/10-injured-nitrogen-explosion-dippin-dots-facility-kentucky/8049493002/

Liquid Nitrogen Leak Prompts Hazmat Situation in Colorado Springs, KKTV 11 News Article, August 26, 2024. https://www.kktv.com/2024/08/26/liquid-nitrogen-leak-prompts-hazmat-situation-colorado-springs/

Florida Mom Warns of Liquid Nitrogen 'Dragon's Breath' Snack After Son's Hospitalized: 'He Could Have Died' By Madeline Farber, Fox News. Published July 31, 2018, 4:46 pm EDT | Updated August 7, 2018, 8: 58 pm EDT. https://www.foxnews.com/health/florida-mom-warns-of-liquidnitrogen-dragons-breath-snack-after-sons-hospitalized-he-could-have-died

Trendy Liquid Nitrogen 'Dragon Breath' Dessert Hospitalizes 14-Year-Old Girl, Fox News Published October 31, 2017, 6: 23 pm EDT | Updated November 1, 2017, 7: 56 am EDT "Nearly Loses Her Thumb Due to Severe Burn". https://www.foxnews.com/food-drink/trendy-liquid-nitrogen-dragon-breath-dessert-hospitalizes-14-year-old-girl

Nitrogen Cocktail Iced After Woman Is Hospitalized in Miami, Reuters May 16, 2014, 2:02 pm CDT. https://www.reuters.com/article/world/us/nitrogen-cocktail-iced-after-woman-is-hospitalized-in-miami-idUSBREA4F0NF/

Additional Resources

FDA Advises Consumers to Avoid Liquid Nitrogen Treats, 12 News by Yaremi Farinas Fri, August 31, 2018 at 7: 30 pm. https://cbs12.com/news/local/fda-advises-consumers-to-avoid-liquid-nitrogen-treats

'Toxic Cocktail' Made With Liquid Nitrogen Sends Florida Woman to the Emergency Room, Medical Daily Published May 14, 2014, 4: 57 pm EDT by Dana Dovey, This article includes a reference to an eighteen-year-old who had to have her stomach removed due to a "Nitro Drink". https://www.medicaldaily.com/toxic-cocktail-made-liquid-nitrogen-sends-florida-woman-emergency-room-282514

Reddit.com has an article about a restaurant spilling liquid nitrogen onto a customer, resulting in painful burns. Please see the following link: https://www.reddit.com/r/legaladvice/comments/16dxi9b/high_end_restaurant_accidentally_spilled_liquid/?rdt=38848

Woman Hospitalized, Loses Gallbladder After Drinking Liquid Nitrogen at Florida Hotel Cryogenic Society of America Archives. https://cryogenicsociety-archive.org/2019/10/23/woman_hospitalized_loses_gallbladder_after_drinking_liquid_nitrogen_at_florida_hotel/

End of Book Quiz

1. What are the two most common causes of nitrogen asphyxiation incidents?
   _____
   _____
   Answer(s):

2. How should you respond if your workmate suddenly and unexpectedly collapses or falls into a process vessel?
   Answer(s):

3. In the event that a loss of containment of nitrogen occurs and you are in a confined space, such as a room or analyser shelter, what actions should you take?
   Answer(s):

4. Why is it important for employees and others working in areas where toxic or asphyxiants are used in the process to participate in periodic drills and exercises?
   Answer(s):

5. Why is the percentage of workers who are killed attempting to rescue a teammate so high?
   Answer(s):

6. Nitrogen is an extreme _____ and _____ hazard. It is present is high concentrations (78%) in the air we breathe. It works by displacing the _____ in the air we breathe; when the nitrogen concentration increases, the _____ concentration decreases.
   Answer(s):

7. When liquid nitrogen evaporates to a vapour, how many times does it expand?
   Answer(s):

8. Liquid nitrogen boils at _____ °F (_____ °C) and contact can cause significant cold burns if it contacts the skin or flesh.
   Answer(s):

Hazards of Nitrogen and Other Inert Gases: How They Can be Safely Managed, First Edition. M. Darryl Yoes.
© 2025 John Wiley & Sons, Inc. Published 2025 by John Wiley & Sons, Inc.

9 Please explain the two primary hazards associated with nitrogen.
   Answer(s):

10 Why is it considered unsafe to connect air to a respirator from a refinery or petrochemical plant compressed air line?
   Answer (s) (select all that apply):
   A The plant air supply line may be mislabelled and may not be air.
   B The plant air supply may be backed up by nitrogen.
   C The plant air system may be unreliable, and the supply may be lost.
   D The facility's air system is too costly to operate.

11 Please name two very good resources for additional information on the hazards of nitrogen.
   Answer(s):

12 For storage tanks, tank trucks, or process vessels that are undergoing a purge with nitrogen, what two things should be placed near the openings or where nitrogen is being vented?
   Answer(s):

13 How much air remains in a typical SCBA when the low-pressure alarms sound?
   Answer(s):

14 Please explain why nitrogen is frequently used as a refrigerant in food processing and the pharmaceutical industries.
   Answer(s):

15. Can you name four of the characteristics of nitrogen?
    Answer(s):

16 Several other gases are also asphyxiants and will displace the oxygen in the air we breathe and can kill.
   Name at least three other gases that are also asphyxiants.
   Answer(s):

17 What is the significant difference between a positive pressure SCBA and a negative pressure SCBA?
   Answer(s):

18 What is an important consideration for the placement of the air compressor intake when the compressor is used to supply breathing air?
   Answer(s):

19 Other asphyxiants include light _____, especially in confined areas or confined spaces.
   Answer(s):

20 What does US OSHA define as entry into a process vessel?

Answer(s):

A OSHA considers entry anytime the head is placed inside a confined space.

B OSHA considers entry to have occurred anytime the hands or feet break the plane of the process vessel and is considered inside the vessel.

C OSHA considers entry to have occurred anytime work is being done inside the vessel and the person is inside the space.

D OSHA considers entry as ensuing work activities in that space and is considered to have occurred as soon as any part of the entrant's body breaks the plane of an opening into the space.

Answer(s):

APPENDIX 1

Answers to the End of Chapter Quizzes

End of Chapter 1 Review Quiz

1. Why is nitrogen an important compound for use in the petroleum refining and petrochemical process industry?

    Answer(s):

    Nitrogen is an inert gas that eliminates oxygen, thereby preventing a reaction with hydrocarbon or other chemicals that may result in a fire or explosion.

2. Nitrogen is used for petroleum and petrochemical tank blanketing to prevent the accumulation of _____ _____ or for products that are sensitive to _____.

    Answer(s):

    flammable   mixtures   air

3. Please explain why liquid nitrogen is frequently used as a refrigerant in food processing and pharmaceutical industries.

    Answer(s):

    The liquid nitrogen temperature is very cold, making it useful as a coolant or refrigerant.

4. What properties make nitrogen more suitable as a gas to pressurize tank cars or railcars to aid in offloading products to storage tanks or to other dispositions?

    Answer(s):

    Nitrogen is inert and will not react with the hydrocarbons or other chemicals that may be in the railcar.

5. Why is nitrogen sometimes used to blanket aircraft fuel tanks?

    Answer(s):

    Nitrogen is inert and displaces oxygen, preventing a flammable mixture from occurring in the storage tanks.

6 Nitrogen makes up about _____% (by volume) of the air that we breathe.

Answer(s): Please select the correct answer(s).

A 52%
B 85%
C 78%
D 26%

Answer(s):

C 78%. Nitrogen makes up 78% (by volume) of the air we breathe. The remainder is oxygen (21%) and a little bit of argon.

7 Can you name four of the characteristics of nitrogen?

Answer(s):

- Colourless
- Odourless
- Non-combustible
- Non-toxic

Nitrogen is a colourless and odourless gas; it is non-combustible and non-toxic. Nitrogen is also slightly lighter than air and is slightly soluble in water.

8 When liquid nitrogen is released into the atmosphere, it will quickly form a white fog by freezing the _____ in the air. It may also freeze anything nearby and can create an oxygen-deficient atmosphere.

Answer(s):

moisture

9 One volume of liquid nitrogen expands to approximately _____ volumes of gas.

Answer(s):

700

10 An unplanned release of liquid nitrogen, for example, from a liquid nitrogen tank, will fill a standard-size room with concentrated nitrogen in _____. People in the room can be quickly overcome by oxygen deprivation, and they can _____.

Answer(s):

seconds   die

11 What mechanism makes nitrogen so dangerous to work with and around?

Answer(s):

In a confined space or where ventilation is poor, nitrogen displaces oxygen from the air we breathe. In other words, when the nitrogen concentration increases, the oxygen concentration decreases.

12 Cold nitrogen is _____ than air; therefore, it will tend to concentrate along the _____ or when in a building or room, along the _____.

Answer(s):

heavier   ground   floor

13 Another significant hazard when working with liquid nitrogen is the possibility of _____ or _____ from contact with the cold liquid.

Answer(s):

burns   frostbite

14 Why is nitrogen purge used before unit start-up for process piping and vessels?

Answer(s): Please select the correct answer(s).

A As a verification that there are no leaks in the piping or vessels before introducing hydrocarbons.
B To eliminate oxygen from the piping or vessels before introducing hydrocarbons.
C To ensure that the piping or vessels will not fail during the start-up process, releasing hydrocarbons into the atmosphere.
D As a verification that there will be no product contamination during the start-up process.

Answer(s):

B Although nitrogen may be used to perform a tightness test prior to start-up, the primary purpose is to eliminate oxygen from the piping or vessels before introducing hydrocarbons to prevent the possibility of creating a flammable mixture in the piping or vessels.

15 How is nitrogen added to plants to help support plant growth?

Answer(s): Please select the correct answer(s).

A Liquid nitrogen is injected into the soil directly below the plants.
B Nitrogen is converted into ammonia or ammonia-based compounds and used as fertilizer to support plant growth.
C Nitrogen is fed to plants by creating a nitrogen atmosphere during the plant's early life.
D Nitrogen is converted into ammonia and the ammonia is provided to the plant in the nursery.

Answer(s):

B Nitrogen is converted into ammonia or ammonia-based compounds and used as fertilizer to support plant growth.

16 Liquid nitrogen is a _____ liquid, that is, it is stored in a pressure vessel specially designed to maintain liquids at extremely low temperatures.

Answer(s): Please select the most correct answer(s).

A condensed
B cryogenic
C mixed
D pure

Answer(s):

B cryogenic

17 When liquid nitrogen is stored, it is at a temperature of _____°F making it very useful as a coolant or refrigerant.

Answer(s): Please select the most correct answer(s).

A 212 °F
B −200 °F
C 32 °F
D −320 °F

Answer(s):

D Liquid nitrogen is at a temperature of −320 °F.

18 Nitrogen is used for tank blanketing to prevent the accumulation of _____ _____ or for products that are sensitive to _____.

Answer(s):

flammable mixtures   air

End of Chapter 2 Review Quiz

1 Several other gases are also asphyxiants and will displace the oxygen in the air we breathe and can kill.

Name four gases (including nitrogen) that are asphyxiants.

Answer(s): Choose nitrogen, plus three of the following that are asphyxiants:

Correct answers are any of the following three.

Argon, Carbon Dioxide, Helium, Neon, Xenon, Light Hydrocarbons.

2 What is the major use of argon in the industry?

Answer(s):

Argon is used to create an inert environment when welding, especially when welding alloy metals.

3 Other asphyxiants include light hydrocarbons, especially in higher concentrations in confined areas or confined spaces. Please name at least three of the lighter hydrocarbons that share these characteristics.

Answer(s) (any three of the following):

Methane, ethane, propane, butane, pentane or any of their olefin cousins such as ethylene, propylene, butylene, etc.

4. What is an evolving use of Helium in petroleum refining and petrochemical plants?

   Answer(s): Name as many as you can.

   A As a pressure test medium for piping circuits and process vessels following a major project or turnaround and before unit start-up.
   B As a lifting gas for lighter-than-air balloons to observe a unit start-up in progress.
   C As a sample gas for the newer version of analysers in a process unit analyser shelter.
   D As a cryogenic coolant in nuclear power plant turbine generators.

   Answer(s):
   A As a pressure test medium for piping circuits and process vessels following a major project or turnaround and before unit start-up.

5. What sets carbon monoxide apart from other gases known to be asphyxiants?

   Answer(s):

   Most asphyxiants work by _____ the oxygen from the air we breathe, and we can be killed by oxygen _____ or _____.

   displacing   deprivation   hypoxia

   Answer(s): Carbon monoxide works by _____ with the transport of oxygen within our body to the internal _____ that depend on oxygen to function.

   Interfering   organs

6. What is common with the classes of light hydrocarbons (methane, ethane, propane, butane, etc.) relative to their potential for causing hypoxia?

   Answer(s):

   When in high concentration, they are all considered to be asphyxiants, especially when they accumulate in areas with low ventilation or confined or restricted spaces.

7. Please name as many other gases that you can that have the same consequences as nitrogen when in higher concentration in confined spaces or other areas of poor ventilation.

   Answer(s) (Please target to name at least four in this category)

   (Correct answers – any four of the following):
   - Argon
   - Carbon dioxide
   - Helium
   - Neon
   - Xenon
   - Krypton
   - Light hydrocarbons (i.e., methane, ethane, propane, butane, pentane and their olefin cousins).

8. Cold nitrogen is _____ than air; therefore, it will tend to concentrate along the _____ or when in a building or room, along the _____.

   Answer(s):

   heavier   ground   floor

9   Another significant hazard when working with liquid nitrogen is the possibility of _____ or _____ from contact with the cold liquid.

Answer(s):

burns   frostbite

10 What is the OSHA Permissible Exposure Limit or Time Weighted Average (TWA) for exposure to Argon?

Answer(s):

A 5 ppm
B 10 ppm
C 25 ppm
D Although argon is well recognized as an asphyxiant, no permissible exposure limit has been established.

Answer(s):

D Although argon is well recognized as an asphyxiant, no permissible exposure limit has been established.

11 What is the primary use of carbon dioxide at petroleum refineries and petrochemical plants?

Answer(s):

Carbon dioxide is typically used as a fire extinguishing agent, for example, in product storage areas or gas turbine generator enclosures.

12 What are the major sources of carbon monoxide in industrial settings?

Answer(s): Answer as many as you can.

A Fired furnaces, heaters or fired steam boilers.
B Automobiles, trucks, railroad engines and other vehicles that burn fossil fuels.
C Portable generators.
D Kilns or calciners.

13 What is the single largest producer of carbon monoxide?

Answer(s):

Vehicle exhaust is the single largest source of carbon monoxide.

14 What is the best resource to find the hazards associated with an inert gas or a toxic chemical?

Answer(s):

The manufacturer's Safety Data Sheet.

15 In this lesson, we learned that light hydrocarbons are very hazardous, including asphyxiants when in higher concentrations, especially in areas of low ventilation or confined spaces.

What remains the primary safety hazard associated with light hydrocarbons?

Answer(s):

Light hydrocarbons are extremely flammable and, under the right circumstances, can be very explosive.

End of Chapter 3 Review Quiz

1. What is the typical oxygen concentration in the air we breathe?

   Answer(s):

   A 23%
   B 40%
   C 12%
   D 19.5%
   E 21%

   Answer(s):

   E 21%. The oxygen concentration in the air we breathe is 21%.

2. When a worker enters an environment where the oxygen level is significantly reduced, how long is it before the effects take place?

   Answer(s):

   A Seconds
   B Minutes
   C Hours
   D Days
   E Weeks

   Answer(s):

   A The effects of a low-oxygen environment can occur in seconds.

3. When a person is overcome due to the lack of oxygen, heart failure and death can occur in about _____ to _____ minutes if they are not _____.

   Answer(s):

   two   four   resuscitated

4. What effects will the workers notice or feel if they enter an oxygen-deprived workspace?

   Answer(s) (Please mark all that apply):

   A Distinct odour or smell
   B Unusual faint feeling
   C Blurred vision
   D No unusual or noticeable effects

   Answer(s):

   D When workers enter an oxygen-deprived environment, there may be no unusual or noticeable effects.

5. What is the maximum oxygen concentration allowed by OSHA in a confined space?

   Answer(s):

   A 19.5%
   B 21%
   C 20%
   D 23.5%

Answer(s):

D The maximum oxygen concentration, as established by US OSHA, is 23.5%.

6 What is the purpose of OSHA establishing a maximum oxygen concentration? Why a maximum?

Answer(s):

Higher concentrations of oxygen can result in regular, everyday materials becoming extremely flammable. Materials such as fabrics can become very flammable and easily ignited.

7 What tactic do welders use to help protect themselves from oxygen deprivation when they are welding using argon to create an inert gas purge around the weld?

Answer(s):

They create internal dams in the piping using tape or pipe plugs to force the argon into the area they are welding, and to help prevent the argon from entering the area where they are working.

8 What device is highly recommended for a welder to use when they are using argon inside an enclosed area such as a pipe or vessel?

Answer(s):

It is highly recommended that welders use an oxygen detector equipped with visual, audible and vibration alarms when the oxygen level is low.

9 Below what concentration of oxygen in breathing air does US OSHA consider the air to be oxygen deficient?

Answer(s):

A 19.5%
B 20%
C 21%
D 23.5%

Answer(s):

A US OSHA considers the air to be oxygen deficient when the concentration is 19.5% or less.

10 Human beings must breathe oxygen to survive and begin to suffer _____ or adverse _____ effects when the oxygen level of their breathing air drops below _____ oxygen.

Answer(s):

hypoxia   health   19.5%

11 Which two US federal agencies are great resources for additional information on the hazards of nitrogen and oxygen deprivation?

Answer(s):

The US Chemical Safety and Hazard Investigation Board (CSB)

The Occupational Safety and Health Administration (OSHA)

End of Chapter 4 Review Quiz

1. At what point in a nitrogen purge procedure is the planning for personnel protection considered?

   Answer(s) (select all that apply):
   A During the pre-job planning.
   B During the job safety analysis.
   C During the pre-job safety checklist.
   D Before any work starts in the field.
   E After the purge is underway.

   Answer(s): A, B, C, D All planning for personnel protection should be included in the pre-job planning, the job safety analysis, the pre-job safety checklist and certainly before any work starts in the field. All planning should be completed before the purge is underway.

2. What are examples of ways personnel are protected against the hazard of oxygen deprivation?

   Answer(s) (select all that apply):
   A Safety warning signs indicating nitrogen purge in progress.
   B Personnel-restricted entry barricades.
   C Personal oxygen monitors with alarms.
   D Continuous gas testing with audible and visual alarms.
   E A self-contained breathing apparatus (SCBA) is required.

   Answer(s): The correct answer is all of the above – These are all methods that can help protect the workers against the hazards of oxygen deprivation.

3. Which US OSHA standard covers the hazards and requirements associated with personnel entering into a confined space?

   Answer(s): US OSHA 29 CFR 1910.146 details the requirements for entry into a permit required confined space.

4. What are the two main personnel safety hazards associated with liquid nitrogen?

   Answer(s): Asphyxiation due to oxygen deprivation and contact with cold liquid results in burns to the skin.

5. How can cold burns be prevented when working around or with liquid nitrogen?

   Answer(s): The simple answer is to avoid contact with the supercold liquids by wearing protective clothing (PPE).

6. At what temperature is liquid nitrogen when it is being stored in a cryogenic storage tank?

   Answer(s): –320 °F or –196 °C

7. What does US OSHA define as entry into a process vessel?

   Answer(s):
   A OSHA considers entry any time the head is placed inside a confined space.
   B OSHA considers entry to have occurred anytime the hands or feet break the plane of the process vessel, which is considered inside the vessel.

C   OSHA considers entry to occur anytime work is done inside the vessel and the person is inside the space.

D   OSHA considers entry as ensuing work activities in that space and is considered to have occurred as soon as any part of the entrant's body breaks the plane of an opening into the space.

Answer(s):

D   OSHA considers entry as ensuing work activities in that space and is considered to have occurred as soon as any part of the entrant's body breaks the plane of an opening into the space.

8   When a nitrogen purge is underway, where is the best place to place the warning signs and personnel restriction barricades?

Answer(s) Very near the vessel opening. It is also recommended that additional signs are located at the vessel access ladders to prevent personnel from entering the platforms where a nitrogen purge is underway.

9   It only takes one breath of _____ nitrogen to make you lose consciousness.

Answer(s): concentrated

10   What steps should a worker take if they observe a team member collapse inside a confined space?

Answer(s): Unless they are a qualified responder, equipped with an SCBA and have a backup person to assist, they should immediately call for help by trained responders. They should then stand by and assist the responders when they arrive.

11   On 5 November 2005, at the Valero Delaware City Refinery, the workers were tasked with installing the top reactor entry piping on the vessel manway. The vessel was under a nitrogen purge to protect the catalyst. What was missing that should have provided the workers with warnings or key information to help protect the workers?

Answer(s) (select all that apply):

A   Safety warning signs indicating nitrogen purge in progress.
B   Personnel-restricted entry barricades.
C   Personal oxygen monitors with alarms.
D   Continuous gas testing with audible and visual alarms.
E   Required use of a self-contained breathing apparatus (SCBA).
F   An accurate work permit indicating nitrogen purge in progress.
G   Protective entry guards or lanyards to prevent workers from falling into the open reactor.

Answer(s):

A–G   All of the above.

12   What is the significant difference between a positive-pressure SCBA and a negative-pressure SCBA?

Answer(s): The difference is that a positive pressure regulator maintains pressure on the mask and prevents atmospheric gases from being drawn into the mask.

13   US OSHA in 29 CFR 1910.146 considers entry to have occurred when _____ part of the body enters the confined space.

Answer(s): any

14 The following lists several potential hazards that may be present when entering a confined space.

Answer(s) (please select all that apply):

A The space may be oxygen deficient.
B The space may contain toxic hazards such as hydrogen sulphide or benzene.
C The space may contain engulfment hazards such as a catalyst.
D The space may contain hydrocarbons that are within the flammable range.
E None of the above.
F All of the above.

Answer(s):

F All of the above

End of Chapter 5 Review Quiz

1 As per the OSHA Confined Space Entry standard, attendant means an _____ stationed outside one or more permit spaces who monitors the authorized _____ and who performs all attendant's duties assigned in the employer's permit space program.

Answer(s):

individual   entrants

2 Permit-required confined space (permit space) means a confined space that has one or more of the following characteristics:
  1 Contains or has the potential to contain a hazardous _____.
  2 Contains a material that has the potential for engulfing an _____.
  3 Has an internal _____ such that an entrant could be _____ or asphyxiated by inwardly converging walls or by a floor that slopes downward and tapers to a smaller cross-section.
  4 Contains any other recognized serious _____ or _____ hazard.

Answer(s):
1 atmosphere
2 entrant
3 configuration trapped
4 safety health

3 What is the purpose of the OSHA interpretation letters?

Answer(s):

The OSHA interpretation letters explain the regulatory requirements and how they apply to particular circumstances.

4 Can or will an OSHA interpretation letter change or create additional employer obligations?

Answer(s):

An OSHA interpretation letter cannot change or create additional employer obligations.

5 How does OSHA define Immediately Dangerous to Life or Health (IDLH) in the Confined Space regulation?

Answer(s):

Immediately dangerous to life or health (IDLH) means any condition that poses an immediate or delayed threat to life that would cause irreversible adverse health effects, or that would interfere with an individual's ability to escape unaided from a permitted space.

6 In the OSHA Confined Space regulation, how do they help protect the worker from accidentally falling into an open manway or other accidental entry into a permit-required confined space or from tools or debris falling onto others who may be working in the confined space?

Answer(s):

The regulation has a requirement as follows:

'When entrance covers are removed, the opening shall be promptly guarded by a railing, temporary cover, or other temporary barrier that will prevent an accidental fall through the opening and that will protect each employee working in the space from foreign objects entering the space'.

7 How does OSHA define a non-permit confined space?

Answer(s):

A non-permit confined space is a space that does not contain or, with respect to atmospheric hazards, has the potential to contain any hazard capable of causing death or serious physical harm.

8 How does US OSHA interpret the requirements for a separate and dedicated rescue service to support emergency rescue if needed during entry into a confined space, or can the entry supervisor, if properly trained and authorized by the employer, act as the standby onsite rescuer to enter the PRCS in the event of an entry rescue, leaving only the attendant at the PRCS entry point?

Answer(s):

OSHA has interpreted the standard to require a separate (either in-house or outside) rescue and emergency service when permitted space entry operations are performed in an IDLH atmosphere.

9 How does US OSHA interpret the confined space standard regarding who is required to wear a rescue harness that is attached to a personnel retrieval system (a hoist or similar)?

Answer(s):

The confined space entry standard (specifically paragraph 29 CFR (1910.146)) (k) (3) focuses on the non-entry rescue of authorized entrants. To facilitate non-entry rescue, a retrieval system must be in place. The paragraph specifically requires, except as explained below, that 'retrieval systems or methods shall be used whenever an authorized entrant enters a permit space'. Thus, OSHA will expect all authorized entrants to wear retrieval devices until it is determined by the employer that a retrieval system presents a greater hazard to the entrant for the space to be entered.

10   In the US OSHA confined space standard, does the term 'body' include all extremities (hands, feet, arms and legs), or does it indicate just the head and torso? What exactly does the standard mean when describing Entry into a Permit-Required Confined Space?

Answer(s):

The term 'body' refers to any part of the anatomy, including all extremities.

11   What specific gas tests are required before entry can be permitted into a permit-required confined space?

Answer(s):

The OSHA regulation 29 CFR (1910.146)(c)(5)(ii)(C) requires, 'Before an employee enters the space, the internal atmosphere shall be tested, with a calibrated direct-reading instrument, for oxygen content, for flammable gases and vapours, and for potential toxic air contaminants, in that order'.

12   Can the employee or their authorized representative view the results of the gas tests before entry into the confined space?

Answer(s):

Yes – Any employee who enters the space, or that employee's authorized representative, shall be provided an opportunity to observe the pre-entry testing required by this paragraph.

13   Is there a requirement for retention of the entry permit, and if so, what period of retention is required, and why is this required?

Answer(s):

Yes – The standard requires that 'the employer shall retain each cancelled entry permit for at least 1 year to facilitate the review of the permit-required confined space program required by paragraph (d)(14) of this section'.

Any problems encountered during an entry operation shall be noted on the pertinent permit so that appropriate revisions to the permit space program can be made.

14   How does US OSHA define an oxygen-deficient atmosphere and an oxygen-enriched atmosphere?

Answer(s):

An oxygen-deficient atmosphere means an atmosphere containing less than 19.5% oxygen by volume.

An oxygen-enriched atmosphere means an atmosphere containing more than 23.5% oxygen by volume.

End of Chapter 6 Review Quiz

1   Why is it not a good idea to connect to a petroleum refinery or petrochemical plant utility air or instrument air supply and use it as breathing air to a sandblasting hood or any other respiratory breathing air system?

Answer(s):

A These systems are not certified as breathing air and have not been tested to ensure they meet the strict requirements for breathing air quality.

B Generally, these systems are backed up by a nitrogen supply in the event of low air pressure. This may subject the user to nitrogen being introduced into a system being used for breathing air, which most often means certain death.

2 What is the primary hazard associated with using blended air or manufactured air as breathing air?

Answer(s):

A Occasionally, the blending process may be off-specification, and the cylinder labelled as breathing air may be filled with primarily nitrogen, which can mean certain death for the end user.

3 In the event blended or manufactured air is used for breathing air, how should it be tested and certified before use in the field?

Answer(s):

A If blended air or manufactured air is used, it should be tested for breathing air quality to ensure the oxygen content is at least 19.5% or higher before it is used.

B Other quality tests for carbon monoxide and other contaminants should also be completed before it is used as breathing air.

4 Compressed oxygen must _____ be used in atmosphere-supplying respirators, including open circuit SCBA's, which have previously used compressed air.

Answer(s):

not

5 Breathing air _____ must be incompatible with outlets for non-respirable plant air or other gas systems to prevent accidental servicing of air-line respirators with non-respirable gases or oxygen.

Also, no _____ substance must be allowed in the breathing air lines.

Answer(s):

couplings

asphyxiating

6 Oil-lubricated compressors can produce _____ _____ if the oil enters the combustion chamber and is ignited.

This problem can be particularly severe in _____ compressors with worn piston rings and cylinders.

Answer(s):

carbon monoxide

older

7 If an oil-lubricated compressor is used to supply breathing air, it must have a _____ _____ or _____ _____ alarm, or both, to monitor carbon monoxide levels.

If only a high-temperature alarm is used, the air from the compressor must be tested for _____ _____ at intervals sufficient to prevent carbon monoxide in the breathing air from exceeding 10 ppm.

Answer(s):

high temperature   carbon monoxide

carbon monoxide

8 What is an important consideration for the placement of the air compressor intake when the compressor is used to supply breathing air?

Answer(s):

The location of the air intake is very important and must be in an uncontaminated area where exhaust gases from nearby vehicles, the internal combustion engine that is powering the compressor itself (if applicable), or other exhaust gases being ventilated from the plant will not be picked up by the compressor air intake.

9 For compressors supplying breathing air, _____ and _____ must be maintained and replaced or refurbished periodically according to the manufacturer's recommendations, and a _____ must be kept at the compressor indicating the most recent change date and the signature of the person authorized by the employer to perform the change.

Answer(s):

sorbent beds and filters   tag

10 What is the maximum percentage of carbon monoxide allowed when testing breathing air for quality?

Answer(s):

A 5%
B 10%
C 15%
D 20%
E 25%

Answer(s):

B 10%

11 Cylinders of purchased breathing air must have a certificate of analysis from the supplier stating that the air meets the requirements for grade D breathing air. The moisture content of the compressed air in the cylinder cannot exceed a dew point of −50 °F (−45.6 °C) at 1 atmosphere pressure.

What is the purpose of this strict moisture requirement?

Answer(s):

This requirement will prevent respirator valves from freezing and blocking the flow of breathing air, which can occur when excess moisture accumulates on the valves.

12 The following list includes the potential hazards and benefits of workers using high-pressure nitrogen to drive their impact guns. Please select those that are clearly hazards and could result in their death under adverse conditions.

Answer(s): Please select those that are hazardous to the workers.

A The nitrogen supply operates at higher pressure, giving the impact guns more power.

B The nitrogen is also drier, making the impact guns more effective.
C The nitrogen may asphyxiate the worker, especially if used in confined spaces or where ventilation is restricted.
D The workers can get the task done faster when using nitrogen to drive the impact guns.
E The use of nitrogen to drive impact guns violates the policy of most companies.

Answer(s):

C The nitrogen may asphyxiate the worker, especially if used in confined spaces or where ventilation is restricted.

End of Chapter 7 Review Quiz

1 At the beginning of this chapter, we discussed some of the more frequent causes of fatalities due to nitrogen asphyxiation and how they can happen. How many of these more frequent causes of nitrogen asphyxiation can you name?

Answer(s): Please list as many of these as you can.

A Plant workers or contractors connected their respirators and/or sandblast hoods to a facility pipeline that was in nitrogen service or to a facility pipeline in air service but, at the time, was backed up by nitrogen.
B Plant workers or contractors entering a storage tank, a truck or other process vessel without verification that the atmosphere would support human life, or without an approved confined space entry permit, or without a respirator supplied with breathing quality air.
C An untrained worker or a worker without proper emergency response and personal protective equipment (PPE) attempting to rescue another worker who has collapsed in or near an open process vessel or a vessel being purged with nitrogen.
D A worker working near an open process vessel that is not gas-free, or a vessel being purged with nitrogen without personal protective equipment (i.e., a respirator supplied with breathing quality air).
E A worker working in an IDLH environment running out of air in their self-contained breathing apparatus (SCBA).
F A worker working in an IDLH environment with the wrong type of respirator (i.e., using an air-purifying respirator instead of breathing air).
G Workers working in a confined space where a section of process tubing, piping or other equipment fails, resulting in the release of nitrogen.
H Workers transporting either pressurized nitrogen or liquid nitrogen cylinders in the cab of a vehicle.

2 What is one thing that each worker can do to help recognize the potential hazard they may face when undertaking an entry into a confined space or about to connect to what they believe to be a supply of breathing air?

Answer(s):

Each worker should pause before immediately reacting to the situation and conduct a 'last minute risk assessment' to think through the tasks they are about to perform and the potential hazards for each task and how those hazards can be mitigated.

3 As a minimum, what should be in place before a worker enters any type of confined space, especially one that may contain a hazard to the worker, such as the potential for lack of oxygen, the presence of hydrocarbons or a chemical, or one that has restricted access and exit or other obstructions or encumbrances?

   Answer(s): Please list as many of these as you can.

   A The entrant and attendant should be properly trained in confined space entry procedures and the use of the breathing apparatus being used.
   B A properly issued and authorized confined space entry permit authorizing his/her entry into the vessel or tank.
   C An authorized and trained attendant to support the person(s) entering the vessel.
   D A properly executed and issued gas test that tested the vessel for oxygen content, hydrocarbon vapours (LEL), and the potential chemicals that may be present in the vessel.
   E Depending on the results of the gas testing, self-contained breathing air equipment such as SCBA or hose line supplied. The breathing air should be pretested to ensure that it is breathing air quality.
   F A personnel rescue plan, potentially with a personnel retrieval harness and hoist or other type of rescue equipment, depending on the rescue plan.
   G Barricades and warning signs to keep all other 'on-lookers' away from where the entry is taking place.

4 No part of the human _____ should be placed inside the confined space until the requirements of the OSHA confined space entry standard are followed.

   Answer(s):

   body

5 What should a worker do if they observe another person fall or enter a confined space without authorization, and if that person has obviously collapsed?

   Answer(s):

   A Immediately call for help from the trained personnel rescue team and standby to aid support if needed.
   B Attempt to hand the person a lifeline or attach a rope to the person to help extract them from the confined space.
   C Immediately put on a SCBA and enter the vessel to remove the person to fresh air.
   D Take action quickly and enter the vessel to rescue their downed buddy.

   Answer(s):

   A Immediately call for help from the trained personnel rescue team and standby to aid support if needed.

6 What are the key features that should be shared by all general unit utility stations that are designed to help make use of utilities safe for all personnel?

   Answer(s): Please name as many as you can.

   A Each utility connection should be clearly labelled and colour-coded to ensure that all workers are familiar with exactly the utility they are connecting to. Each of the four main utilities—air, steam, water, and nitrogen—should be clearly labelled and colour-coded, with a different colour for each utility.

B Each of these utilities should also be equipped with separate and unique quick connections, a different connection for each utility. This design should make it impossible to connect one utility to another.

C Each connection should also be equipped with a check valve or non-return valve to prevent backflow from the utility into the process.

7 How does OSHA attempt to protect a worker from being overcome by nitrogen (oxygen deprivation) or other dangerous vapours and falling into an opening in a process vessel, or from workers inside the space from being struck by falling tools or debris?

Answer(s):

The OSHA regulation 29 CFR (1910.146)(c)(5)(ii)(B) requires 'When entrance covers are removed, the opening shall be promptly guarded by a railing, temporary cover, or other temporary barrier that will prevent an accidental fall through the opening and that will protect each employee working in the space from foreign objects entering the space'.

8 What is unsafe about transporting pressurized nitrogen cylinders or other compressed gas cylinders in the cab of a pickup truck or any other vehicle enclosure, such as a car trunk?

Answer(s):

Pressurized cylinders have been known to leak and can displace the oxygen inside the enclosed vehicle and can result in the deaths of the driver and passengers.

Since nitrogen has no odour or other indication of the release, there is no warning to the occupants, and they die.

9 Why is it that an air-purifying respirator will not protect a worker against oxygen deprivation in the event it is used in a nitrogen environment?

Answer(s):

An air-purifying respirator can help remove particulate matter or dust but does not provide the oxygen needed to support life. Nitrogen displaces oxygen from the environment, leading to hypoxia or asphyxiation of anyone exposed. Only a respirator with certified breathing air should be used when exposure to nitrogen or any asphyxiant may be possible.

10 What is the hazard of a worker who is working near an open process vessel that is not gas-free, or a vessel being purged with nitrogen without personal protective equipment (i.e., a respirator supplied with breathing-quality air.)?

Answer(s):

There have been several cases where workers who were working near an open vessel or a vessel under nitrogen purge, without PPE or breathing air equipment, have been overcome by oxygen deprivation and have fallen into the vessel and died.

End of Chapter 8 Review Quiz

1 What is unsafe about craftsmen using nitrogen to drive impact wrenches?

Answer(s):

A The nitrogen exhausts almost directly into the faces of the craftsmen. Nitrogen can easily displace the oxygen needed for respiration, and the craftsmen can be injured or killed by oxygen deprivation.

B Impact wrenches can also be taken into a confined space or other areas with low ventilation and suffer the same effects.

2 What is unique about the typical design for potable water systems?

Answer(s):

Potable water systems are designed and constructed to ensure there are no direct connections between them and other process piping or vessels. This prevents potable water from being contaminated by chemical or hydrocarbon-containing systems.

3 What is a common way for petroleum refining or petrochemical facilities to identify utility systems such as utility air, steam, utility water and nitrogen?

Answer(s):

Utility systems should each be painted a distinctive colour and labelled as to the utility.

4 Is it common to see or have multiple types of quick connections installed on the same utility piping system?

Answer(s):

No – each utility system should be provided with a unique quick connection specific to that utility. For example, air systems should have one type of connection, and that connection should be on all air systems. The same is true for the other three most common utility systems; they should each have a dedicated quick connection specific to that utility.

5 Why is it important to have a specific quick connection for each utility that is not compatible with connections used for other utility lines?

Answer(s): Please select the most correct answer.

A So that an operator can easily identify the proper utility system before connecting it to process equipment.

B To prevent cross-connecting one utility to another utility and contaminating the system.

C To ensure that operators don't connect to the wrong utility.

D To make it easy for the operators to keep the utility systems well organized.

Answer:

B To prevent cross-connecting one utility to another utility and contaminating the system.

6 Is it okay for operators to fabricate and use a utility cheater connection to allow connecting one utility to a different utility?

Answer(s):

No – utility cheater systems typically violate the policy of keeping the utility systems segregated and can result in the contamination of one or more utility systems.

7 What is the purpose of the check valve or non-return valve, which is typically installed on utility stations?

Answer(s):

To prevent the chemicals or hydrocarbons from a connection made to the process from backflowing into the utility, thereby contaminating the utility system.

8   What is an important aspect of utility hoses to ensure that they are properly rated and serviceable?

Answer(s):

Each utility hose should be properly labelled, indicating the allowable service, the rated pressure and rated temperature. The hose should also have a current inspection tag indicating that it has been pressure tested and is serviceable.

9   What is the purpose of a whip check device when installed on a high-pressure steam hose?

Answer(s):

The whip check is a safety device that, when properly installed, can prevent a flailing hose in the event of a coupling or connection failure.

10 When the whip check is properly installed on a steam hose, should the cable loops be close to the coupling or as far from the coupling on each side as possible?

Answer(s): As indicated in the image, the cable loops should be stretched tight and located as far from the coupling as possible.

11 Is it common and acceptable to have two different utility quick connections available on a common utility supply line?

Answer(s): This should never be acceptable. This leads to improper use of utility hoses and could lead to a connection to the wrong utility.

End of Chapter 9 Review Quiz

1   How can a catalyst crust be identified in a reactor about to undergo an inert entry to replace the catalyst?

Answer(s):

Conduct a differential pressure survey across the reactor. If pressure is below the catalyst, there is likely a crust on top of it.

2   What is the hazard of a crust on top of the catalyst when planning an inert entry into a reactor?

Answer(s):

The pressure can build up and rupture the catalyst bed, releasing a violent force that can injure the inert entry contractors.

3   What is the significant hazard associated with conducting inert entry into a refining reactor?

Answer(s):

The workers are entering an inert reactor completely void of oxygen. They are working in an environment that is immediately dangerous to life and health (IDLH).

4   Why is it important to inert the reactor vessel with nitrogen or another inert when opening the reactor to the atmosphere?

Answer(s):

The catalyst is pyrophoric and will self-ignite when exposed to air.

5   What is the maximum oxygen concentration allowable in the reactor when conducting an inert entry into the reactor?

What is the hazard we are addressing with this very low oxygen concentration?

Answer(s):

The atmosphere inside the reactor is typically 99% nitrogen with an oxygen concentration of about 1%. The typical maximum is 3% and if the oxygen concentration exceeds 5% the workers should be immediately evacuated from the reactor.

The very low oxygen concentration prevents pyrophoric ignition of the catalyst.

6   How do we know that the quality of the compressed air meets the OSHA standards for grade D breathing air?

Answer(s):

We have received the certificate of analysis from the supplier for each cylinder of breathing air, indicating it is approved for breathing air quality. We have also followed up and sampled each cylinder to verify that each cylinder meets the specifications for oxygen content (19.5–23%).

7   Why is it important to have a preplanned rescue plan and trained responders on-site during the vessel entry?

Answer(s):

In an emergency situation involving rescue, there will not be time to call for help from outside the facility. The dedicated responders are trained professionals in this type of rescue and have the specialized equipment required.

8   Why are specialized clam-shell or anti-panic helmets required for entrants when conducting inert entry?

Answer(s):

The clamshell or anti-panic helmets are designed to prevent workers from attempting to remove them, for example, if they have a panic attack while in an inert environment.

9   What is the maximum LEL (flammables) allowable before entry into a reactor vessel when undergoing inert entry for catalyst replacement?

Answer(s): Please select the most correct answer.

A   2%
B   5%
C   10%
D   15%
E   25%

Correct answer:

C 10%

10 When the test for carbon monoxide is positive, what chemical may be present in the reactor?

Answer(s):

Carbonyls, which may be toxic to the workers.

11 During an inert entry, how many breathing air sources are available to those working in the reactor?

Can you name the breathing air systems?

Answer(s):

Four sources of breathing air are available to the workers: the primary and backup air systems fed directly to their helmets, the 5-minute escape pack and the Emergency Exit Pack.

12 What special PPE should be worn by the workers who are testing the reactor for LEL before permitting entry into the inert atmosphere?

Answer(s):

A self-contained breathing apparatus (SCBA) and fall protection.

13 The primary nitrogen supply during vessel entry should be a _____ system and from a reliable source. A backup nitrogen supply that is completed _____ of the primary supply should be readily available and also reliable. Entry is only allowed when using the _____ source.

Answer(s):

segregated   independent   primary

14 What is the maximum oxygen allowed in the nitrogen purging system?

Answer(s):

Typically 1% with a 3% maximum concentration.

15 Before the entry, the inert entry contractor must submit a detailed _____ _____ _____ to the facility for approval/endorsement.

Answer(s):

Site-Specific Safety Plan

16 At most companies, are company employees allowed to suit up and enter a vessel with an inert atmosphere?

Answer(s):

Inert entry is highly specialized work and requires substantial training and qualification. Also, highly specialized tools and equipment are required. Therefore, most companies do not allow employees to enter confined spaces with an atmosphere that is immediately dangerous to life and health (IDLH).

17 During most inert activity tasks, who is monitoring the breathing air pressures and breathing air rates and how is it monitored?

Answer(s):

The breathing air pressures and breathing rates are monitored by an Inert Entry Specialist from a breathing air console. This person is in contact with the entrants during the entire process.

End of Chapter 10 Review Quiz

1 Why is carbon capture so important?

Answer(s):

Carbon Capture, Use and Storage is an environmental strategy with the goal to capture carbon dioxide from the atmosphere and ongoing operations, as well as store or 'sequester' the recovered gas to prevent it from contributing to greenhouse gases, which can harm the environment.

2 Which inert gas is carbon capture focused on to remove it from our environment and help prevent the formation of greenhouse gases?

Answer(s):

Carbon dioxide (CO_2)

3 CO_2 is an inert gas, and it is an asphyxiant, especially in high concentrations or in enclosed areas or areas where there is little or no air circulation. People have been killed by exposure to CO_2 due to the displacement of _____ from their environment.

Answer(s):

Oxygen

4 When building new projects or modifying others to extract and process CO_2, the hazards must be considered during the _____, _____, _____ and _____ operational phases of the modified or new equipment being brought online.

Answer(s):

design   construction   start-up   online

5 CO_2 does not naturally exist in a liquid state. However, it can be liquefied, but only under specific conditions of high _____ and low _____.

Answer(s):

Pressure   temperature

6 According to the Environmental Protection Agency (EPA), in 2022, CO_2 accounted for _____% of all US greenhouse gas emissions from human activities.

Answer(s):

A 10

B 25

C 65
D 80
E 90

Answer(s):

d 80%

7 What is a leading alternative use of carbon dioxide after it is captured from the environment?

Answer(s):

One of the most promising is the use of CO_2 in the enhanced recovery of hydrocarbons from underground wells.

8 What are the five current processing strategies or technologies being explored for carbon capture?

Answer(s):

Solvent extraction   Sorbents   Membranes   Cryogenics   Oxy combustion

9 The most common and by far the most efficient and safest way to transport carbon dioxide is by _____.

Answer(s):

pipeline

10 What is meant by the term sequestration?

Answer(s):

Storage of CO_2 in deep rock crevices or formations such as depleted oil and gas wells, coal beds, saline aquifers or similar caverns or structures. This prevents the formation of greenhouse gases in the atmosphere.

End of Chapter 11 Review Quiz

1 What is by far the most common asphyxiant in general industry, the one that results in more deaths than most of the others combined? Why is this?

Answer(s):

Nitrogen is by far the most common cause of asphyxiation than all the others combined.

Nitrogen is the most readily available, as 78% of the air we breathe is nitrogen. It is also the most frequently used inert gas in industry.

2 Following nitrogen, what is most likely the second most frequent gas that is involved in employee or worker fatalities? Why?

Answer(s):

More than likely this would be argon gas.

Argon is frequently used in welding operations and is also known as a deadly asphyxiant because it can displace oxygen when used in a confined space or where there is restricted ventilation.

3 What was the cause of the tragic incident that occurred at the Foundation Foods poultry processing facility on 28 January 2021?

How many people died in this tragic accident?

Answer(s):

A level controller failed on an immersion freezer, resulting in an overflow of liquid nitrogen into the surrounding processing room.

Six employees were killed in this tragic incident.

4 At the Valero Delaware City Refinery, what information did the workers have that informed them that the reactor was being purged with nitrogen?

Answer(s):

The workers were not provided with information that the reactor was being purged with nitrogen and that nitrogen was flowing from the top manway.

- The sign indicated 'confined space – no entry'.
- The work permit indicated 'nitrogen purge N/A'.
- No warning was provided about the nitrogen purge until after the two fatalities when the sign was replaced, indicating a nitrogen purge was underway.

5 How could the incident at Foundation Foods have been prevented, or at least how could the consequences and loss of life have been prevented?

Answer(s): Please list all that you can.

- The level control device should have been equipped with a backup level control system and alarms to indicate that the level was higher than normal.
- Provide all employees with training and procedures on the hazards of working with liquid nitrogen and how to respond in the event of a nitrogen release.
- Provide specialized equipment and personnel protective equipment designed to detect and protect personnel from working in oxygen-deficient areas. This includes personnel oxygen monitors with alarms and self-contained breathing apparatus.
- The processing room should have been equipped with forced fresh air ventilation to displace nitrogen in the event of a leak or release.
- The rooms containing or using nitrogen should be equipped with low-oxygen alarms to indicate the accumulation of nitrogen or the absence of oxygen in the room.
- Conduct regular planned employee drills and exercises to practice response to a loss of containment of liquid nitrogen.
- Implement a dedicated emergency response team and plan to minimize response by non-trained personnel.

6 What is the most likely outcome if one worker enters a confined space and is overcome by the lack of oxygen and his workmate decides to enter the confined space to attempt a rescue (without training and without specialized rescue equipment)?

Answer(s):

Based on industry experience, it is most likely that if a worker enters to confined space to rescue his teammate without specialized rescue training and specialized rescue equipment, the attempted rescue will result in a second fatality.

7 As defined by the OSHA regulation for confined space entry, what must be in place before an entrant can enter a confined space if it meets the definition of a permit-required confined space?

Answer(s): Please list as many of the requirements as you can.

- If employees are to enter a permit required confined space, the employer must develop and implement a written permit space program that fully complies with the regulation.
- The space shall be prepared by purging, inerting, flushing or ventilating as necessary to eliminate or control atmospheric hazards.
- Danger signs must be posted to inform employees and others of the potential hazards and to keep people at a safe distance from the openings. The regulation provides examples of the wording to be used.
- When openings or covers are removed, the openings shall be protected by guards, railing, a temporary cover or other barrier to prevent an accidental fall into the opening.
- A gas test confirming there is adequate oxygen (a minimum of 19.5%), that toxics are identified and are within the limits approved by the OSHA standards, and the lower explosive limit is no more than 10%. The atmosphere within the space shall be periodically tested as necessary to ensure the ventilation prevents a hazardous atmosphere. If a hazardous atmosphere is detected, employees shall leave the space immediately.
- A pre-entry written certification shall be developed that includes the date, the space location, and the signature of the person providing the certification before entry into the permit-required space.
- Provide at least one attendant outside the confined space and in communication with all those in the confined space. The attendant must be capable of calling for help or support in an emergency.
- The entry permit must be retained for at least one year to facilitate the review of the permit-required confined space program.

8 US OSHA has two definitions for confined space: a non-permit confined space and a permit-required confined space. What is the major difference between the two definitions?

Answer(s):

- Non-permit confined space means a confined space that does not contain or, with respect to atmospheric hazards, have the potential to contain any hazard capable of causing death or serious physical harm.
- Permit-required confined space (permit space) means a confined space that has one or more of the following characteristics:
  1 Contains or has the potential to contain a hazardous atmosphere,
  2 Contains a material that has the potential for engulfing an entrant,
  3 Has an internal configuration such that an entrant could be trapped or asphyxiated by inwardly converging walls or by a floor, which slopes downward and tapers to a smaller cross-section; or
  4 Contains any other recognized serious safety or health hazard.

9 We know that nitrogen and argon have no odour or other way we can be made aware of their presence, and their ability to displace oxygen makes them extremely dangerous, especially in a confined space or where there is restricted or no ventilation.

Is this also true for carbon dioxide (CO_2 or dry ice)?

Answer(s):

Yes – CO_2 (or dry ice, which is carbon dioxide in the solid state) does not have an odour and can also displace the oxygen in a confined space or area where there is restricted or no ventilation. CO_2 can be just as deadly as nitrogen or argon.

End of Chapter 12 Review Quiz

1 What is each facility required to have in place before an entry into permit-required confined space can be executed?

Answer(s):

Each facility must have a plan and procedure that includes all of the elements and requirements as outlined in the OSHA regulation 29 CFR (1910.146).

2 Please provide the primary reasons why policies or detailed procedures should be in place that prevent workers from connecting an airline-supplied respirator to a facility pipeline for use as breathing air.

Answer(s):

A Pipelines have been identified incorrectly, and workers have been subjected to breathing nitrogen while thinking they were connected to a pipeline containing air.
B Some air pipelines are backed up by nitrogen and will supply nitrogen in the event of a compressor failure or any other cause of low air pressure.
C Many workers have died as a result of connecting to a pipeline system they believed to be breathing air but was another utility, primarily nitrogen.

3 Always ensure that workers follow the process to ensure the atmosphere is safe before _____ any process storage tank or vessel. This means that before entering a vessel verification that all hydrocarbon sources are effectively isolated, all hazardous material has been removed, the atmosphere has been tested for hydrocarbons (lower explosive limit), toxins and _____ deficiency, and an entry permit has been issued.

Answer(s)

entering   oxygen

4 Why is it important for workers and others to understand that they should never transport nitrogen cylinders, or pressurized cylinders of any type, in the cab of a pickup truck or any other enclosed vehicle, even in the trunk of a vehicle?

What makes this an extreme hazard for those in the vehicle with the pressurized cylinders?

Answer(s):

Cylinder isolation valves can leak, and nitrogen and other inert gases can displace the oxygen inside the vehicle enclosure.

This is an extreme hazard for those in the vehicles because if the cylinder leaks, there is no odour and no warning that this is occurring, and people have died transporting cylinders inside vehicles.

5 How should the space be verified to be free of hazards before an entry into a permit-required confined space can be attempted?

Answer(s):

All hydrocarbon sources are effectively isolated, all hazardous material has been removed, the atmosphere has been tested for hydrocarbons (LEL), toxins and oxygen deficiency, and an entry permit has been issued.

6 How can workers be made aware of alarms and the actions to take in the event an alarm is activated?

How can this be enforced on a routine basis?

Answer(s):

Employees and others should be trained on the proper response procedures, including awareness of each alarm's visual and/or audible signals and actions to take if an alarm is activated.

This can be enforced by conducting routine emergency drills and exercises to reinforce the proper response to abnormal/alarm conditions.

7 What actions should be taken if an entry into a permit required space is ongoing and an alarm sounds or other similar unexpected event or condition arises?

Answer(s):

All work should cease, workers should immediately exit the confined space, and the site should be reassessed for potential hazards. This includes conducting another gas test for oxygen, toxins and flammables to ensure the space is safe to re-enter.

8 Is it possible for low oxygen conditions to exist outside of confined spaces, especially when nitrogen purge is underway?

How should we respond when this is recognized?

Answer(s):

Yes – Oxygen-deficient atmospheres can be dynamic and can exist outside of confined spaces.

Analyse all potential facility conditions and implement protections with the hierarchy of controls.

9 What actions should workers take if they see an associate collapse while working inside a permit-required confined space?

Should they attempt a rescue?

Answer(s):

Ensure that workers know that they are not to perform an attempted rescue of an associate unless they have received specialized rescue training and are equipped with specialized rescue equipment, including breathing air.

Instead, they should immediately call for help from responders who are trained in rescue techniques and who have specialized rescue equipment.

10 What are the important design parameters for emergency gas detection alarms?

Answer(s): Answer all you can.

A Alarm lights and audible sounds should be located where all workers can see or hear the alarms if/when they are activated.
B Adequate signs and posters should be placed where all workers can see and understand what warnings to expect if the alarms are activated.
C Signs or posters should specify the action to be taken in event the alarms are activated.
D Alarms should be tested on a regular basis to ensure they are fully functional.
E Alarms should be incorporated into worker training and drills.

End of Chapter 13 Review Quiz

1 We learned earlier that two of the main hazards of liquid nitrogen are the potential for cold burns upon contact and asphyxiation from oxygen displacement. What is the third serious hazard with liquid nitrogen?

Answer(s):

Ingestion of even a very small portion of liquid nitrogen

2 Liquid nitrogen is typically supercold (_____°F) and can cause catastrophic burns to the digestive tract. When it vaporizes, liquid nitrogen expands _____ times, resulting in extreme internal injuries.

Answer(s):

−320 °F

700

3 One of the most tragic nitrogen incidents to occur in recent times happened at the _____ _____ _____, where six people died, four of whom were attempting rescue.

Answer(s):

Foundation Food Group

4 An injury or illness resulting from ingesting liquid nitrogen while eating nitrogenated ice cream on the job is not reportable to OSHA.

Answer(s): Please mark the correct answer.

True

False

Answer(s): This statement is true, OSHA considers eating on the job to be personal activity and not work-related. However, if the ice cream was provided by the employer or was contaminated onsite, it is considered work-related and would be recordable.

5 Workers who handle liquid nitrogen or work near equipment containing liquid nitrogen should be properly _____ and wear proper _____ _____ _____.

Answer(s):

trained

personal protective equipment (PPE).

6 A leak in a liquid nitrogen storage tank in an enclosed space like an ice cream shop can expose people to _____ _____ or other safety concerns.

Answer(s):

nitrogen asphyxiation

APPENDIX 2

Answers to the End of Book Quiz

End of Book Quiz (With Answers)

1. What are the two most common causes of nitrogen asphyxiation incidents?

   Answer(s):

   Connecting a breathing air respirator to a facility nitrogen pipeline, believing it to be an air line.

   Entering a storage tank, truck compartment or other process vessel without verification that the atmosphere would support human life, without an entry permit, or without a fresh air-supplied respirator.

2. How should you respond if your workmate suddenly and unexpectedly collapses or falls into a process vessel?

   Answer(s):

   We should always call for help and avoid rushing in to provide assistance unless we are trained as a responder and are properly equipped with the proper personnel protective equipment.

3. In the event that a loss of containment of nitrogen occurs and you are in a confined space, such as a room or analyser shelter, what actions should you take?

   Answer(s):

   Immediately leave the area and go outside or into fresh air and quickly tell anyone else in the enclosed area who may be exposed to leave. Immediately call for emergency response and prevent anyone from entering.

4. Why is it important for employees and others working in areas where toxic or asphyxiants are used in the process to participate in periodic drills and exercises?

   Answer(s):

   This is important for workers to understand what actions to take and how and where to respond in the event of a loss of containment.

5 Why is the percentage of workers who are killed attempting to rescue a teammate so high?

Answer(s):

It is a natural human response to want to rescue and help a teammate who has collapsed while working in a hazardous environment. Unless you are a trained responder and have the appropriate protective equipment, this must be resisted and allow our training to take over. We should immediately call for help and then stand out of harm's way and direct the responders to the site when they arrive.

6 Nitrogen is an extreme _____ and _____ hazard. It is present in high concentrations (78%) in the air we breathe. It works by displacing the _____ in the air we breathe; when the nitrogen concentration increases, the _____ concentration decreases.

Answer(s):

inhalation and asphyxiating   oxygen   oxygen

7 When liquid nitrogen evaporates to a vapour, how many times does it expand?

Answer(s):

Liquid nitrogen expands 700 times when it evaporates to a vapour.

8 Liquid nitrogen boils at _____ °F (_____ °C) and contact can cause significant cold burns if it contacts the skin or flesh.

Answer(s):

−320 °F or −196 °C

9 Please explain the two primary hazards associated with nitrogen.

Answer(s):

In either form, as a liquid or as a gas, nitrogen's deadly characteristic is that it quickly displaces oxygen from the environment. Many people have been killed by oxygen deprivation when nitrogen displaces oxygen.

The second hazard is associated with liquid nitrogen. Liquid nitrogen is a cryogenic liquid; that is, it is stored in a pressure vessel specially designed to maintain liquids at extremely low temperatures. Liquid nitrogen is stored at a temperature of −320 °F (−195 °C) and is an extreme risk of causing cold burns if it contacts the skin or flesh.

10 Why is it considered unsafe to connect air to a respirator to a refinery or petrochemical plant compressed air line?

Answer(s) (select all that apply):

A The plant air supply line may be mislabelled and may not be air.
B The plant air supply may be backed up by nitrogen.
C The plant air system may be unreliable, and the supply may be lost.
D The facility's air system is too costly to operate.

Answer(s):

A The plant air supply line may be mislabelled and may not be air.
B The plant air supply may be backed up by nitrogen.
C The plant air system may be unreliable, and the supply may be lost.

11 Please name two very good resources for additional information on the hazards of nitrogen.
Answer(s):
The US Chemical Safety and Hazard Investigation Board (CSB).
The US Occupational Safety and Health Administration (OSHA).

12 For storage tanks, tank trucks or process vessels that are undergoing a purge with nitrogen, what two things should be placed near the openings or where nitrogen is being vented?
Answer(s):
Warning signs notifying people that they are approaching an area that may be oxygen deficient and is a hazardous area.
A barricade to warn people that a nitrogen purge is in progress, the area may be oxygen deficient and they should not enter.

13 How much air remains in a typical SCBA when the low-pressure alarms sound?
Answer(s):
About 25–33% of the original air inventory remains when the alarm sounds. For a thirty-minute cylinder, this means that about 7.5 minutes to almost 10 minutes remain.

14 Please explain why nitrogen is frequently used as a refrigerant in food processing and the pharmaceutical industries.
Answer(s):
Liquid nitrogen is stored in specially designed pressure vessels designed to protect the super low temperature of the stored cryogenic liquid. When it is withdrawn, the temperature is −320 °F (−195 °C), making it very useful as a coolant or refrigerant.

15 Can you name four of the characteristics of nitrogen?
Answer(s):
Nitrogen is a colourless and odourless gas; it is non-combustible and non-toxic. Nitrogen is slightly lighter than air and is slightly soluble in water.

16 Several other gases are also asphyxiants and will displace the oxygen in the air we breathe and can kill.
Name at least three other gases that are also asphyxiants.
Answer(s) (any three of the following):
Argon, carbon dioxide, helium, neon, xenon, light hydrocarbons.

17 What is the significant difference between a positive pressure SCBA and a negative pressure SCBA?

Answer(s):

A positive pressure SCBA delivers pressurized air slightly above ambient air pressure. This prevents smoke or vapours from entering the mask, even if the mask has minor leaks at the seal with the individual's face.

A negative pressure SCBA depends on the user to inhale creating a negative pressure inside the mask before supplying air for the user to breathe. This can allow smoke or vapours to enter the mask when minor leaks occur at the face seal to the mask.

18 What is an important consideration for the placement of the air compressor intake when the compressor is used to supply breathing air?

Answer(s):

The location of the air intake is very important and must be in an uncontaminated area where exhaust gases from nearby vehicles, the internal combustion engine that is powering the compressor itself (if applicable), or other exhaust gases being ventilated from the plant will not be picked up by the compressor air intake.

19 Other asphyxiants include light _____, especially in confined areas or confined spaces.

Answer(s):

hydrocarbons

20 What does US OSHA define as entry into a process vessel?

Answer(s):

A OSHA considers entry anytime the head is placed inside the confined space.
B OSHA considers entry to have occurred anytime the hands or feet break the plane of the process vessel and are considered inside the vessel.
C OSHA considers entry to have occurred anytime work is being done inside the vessel and the person is inside the space.
D OSHA considers entry as ensuing work activities in that space and is considered to have occurred as soon as any part of the entrant's body breaks the plane of an opening into the space.

Answer(s):

D OSHA considers entry as ensuing work activities in that space and is considered to have occurred as soon as any part of the entrant's body breaks the plane of an opening into the space.

APPENDIX 3

Documented Incidents Involving Nitrogen Resulting in Fatalities or Serious Injury

March 1984 OSHA Report - Ball Aerospace Systems Division (1 Fatality)

ON MARCH 31, 1984, BETWEEN 9:45 AM AND 10:15 AM, EMPLOYEE #1 ENTERED A THERMOTRON REFRIGERATION UNIT (7′×7′ ×7′10″) WHICH WAS BEING USED AS AN ENVIRONMENTAL TEST CHAMBER. THE CHAMBER HAD BEEN FILLED WITH NITROGEN FOR COOLING PURPOSES. EMPLOYER #1 ENTERED THE CHAMBER TO INSTALL A VIDEO MONITOR, WHICH NEEDED TO BE TESTED. THE CHAMBER WAS NEITHER PURGED WITH OXYGEN PRIOR TO ENTRANCE NOR TESTED FOR OXYGEN DEFICIENCY. EMPLOYEE #1 WAS FOUND IN THE CHAMBER BY A CO-WORKER AT APPROXIMATELY 10:20 AM. THE CO-WORKER PULLED EMPLOYEE #1 FORM THE CHAMBER AND BEGAN ADMINISTERING FIRST AID. EMPLOYEE #1 WAS TRANSPORTED TO BOULDER COMMUNITY HOSPITAL BY AMBULANCE AND WAS PRONOUNCED DEAD AT 11:17 AM.

https://www.osha.gov/ords/imis/accidentsearch.accident_detail?id=14506000

January 1985 OSHA Report - Texaco Refining & Marketing Inc (1 Fatality)

At approximately 11:00 a.m. on December 5, 1984, Employee #1 was installing a plate in a reactor vessel when he collapsed. He died on December 20, 1984. Tests conducted by the company indicated that oxygen levels in the reactor had been at 10 to 15 percent due to nitrogen displacement.

https://www.osha.gov/ords/imis/accidentsearch.accident_detail?id=14412985

February 1985 OSHA Report - Ibm Corporation (1 Fatality)

ON FEBRUARUY 5, 1985, EMPLOYEE #1 WAS ASSIGNED TO PERFORM AN OPERATION INVOLVING REINTRODUCTION OF TONER IN A BLENDER ROOM OF THE TOWER MANUFACTURING DEPARTMENT. EMPLOYEE #1 WAS LAST SEEN ALIVE AT APPROXIMATELY 8:15 PM. AT APPROXIMATELY 8:35 PM, TWO OTHER EMPLOYEES ENTERED THE BLENDER ROOM AND FOUND EMPLOYEE #1 UNCONSCIOUS ON THE FLOOR. EMPLOYEE #1 WAS WEARING A YELLOW, FULL-BODY COVERALL WITH ATTACHED HOOD AND FACESHIELD. THE COVERALL WAS MANUFACTURED BY DURAFAB, INC., STYLE 2130Z. IT WAS THE "BESTOSHIELD SUIT," DESCRIBED AS A RESPIRATOR HOODED

COVERALL. AN AIR HOSE WAS TAPED INTO THE RESPIRATOR, AND THE OTHER END WAS CONNECTED TO A NITROGEN GAS LINE WITH A QUICK-CONNECT FITTING. THE COUNTY CORONER DETERMINED THE CAUSE OF DEATH TO BE ASPHYXIATION BY THE INHALATION OF NITROGEN GAS. THE QUICK-CONNECT FITTINGS IN THE BLENDER ROOM WERE THE SAME SIZE AND TYPE FOR THE NITROGEN GAS LINES AS THEY WERE FOR THE COMPRESSED AIR GAS LINES. IN ADDITION, THE NITROGEN AND COMPRESSED AIR GAS LINES WERE OF THE SAME COLOR AND, EXCLUDING THE IDENTIFYING NAME, OF SIMILAR LABELING.

https://www.osha.gov/ords/imis/accidentsearch.accident_detail?id=14507347

May 1985 OSHA Report - Chaparral Steel (1 Fatality)

EMPLOYEE #1 WORKED ON A PLATFORM AS A POURER IN A STEEL FOUNDRY. HIS JOB WAS TO CONTROL THE LEVEL OF MOLTEN METAL IN A TUNDISH SEVERAL FEET BELOW. HE WAS ASPHYXIATED WHEN HE DONNED A 3 M BRAND WHITECAP W-5002 SUPPLIED AIR RESPIRATOR THAT HAD BEEN CONNECTED TO NITROGEN RATHER THAN COMPRESSED AIR. AN UNMARKED PLANT NITROGEN LINE HAD BEEN BROUGHT UP TO THE PLATFORM FOR USE BY THE RESPIRATOR WEARER. NO OTHER GASES WERE USED ON THE PLATFORM.

https://www.osha.gov/ords/imis/accidentsearch.accident_detail?id=14465389

July 1985 OSHA Report - Roy Hunt Inc / Koch Refining Company (1 Fatality)

A PETROCHEMICAL REFINERY WAS UNDERGOING TURNAROUND. A CONTRACTOR CREW HAD BEEN ASSIGNED THE JOB OF SANDBLASTING THE INSIDE SURFACES OF A REACTOR VESSEL (CONFINED SPACE). ALTHOUGH UNWRITTEN COMPANY POLICY CALLED FOR THE USE OF AIR COMPRESSORS FOR SUPPLYING RESPIRATOR AIR, THIS CREW, WITH THE SUPERVISOR'S KNOWLEDGE, HAD PREVIOUSLY USED PLANT AIR TO SUPPLY BREATHING AIR. THE CREW MISTAKENLY HOOKED UP AN AIR FILTER TO AN UNLABELED NITROGEN LINE (ONLY THE SHUTOFF VALVE WAS LABELED) USED BY THE REFINEREY FOR PURGING CONFINED SPACES. PLANT NITROGEN LINES AND AIR LINES WERE IDENTICAL AND BOTH HAVE CROW'S FOOT COUPLINGS COMPATIBLE WITH THE COUPLING ON THE AIR FILTER. EMPLOYEE #1 CLIMBED TO THE BOTTOM OF THE VESSEL, PLACED THE SANDBLAST RESPIRATOR HOOD ON HIS HEAD AND BEGAN SANDBLASTING. HE WAS ASPHYXIATED WITHIN FIVE MINUTE'S TIME. ATTEMPTS TO REVIVE EMPLOYEE # WITH OXYGEN FAILED.

https://www.osha.gov/ords/imis/accidentsearch.accident_detail?id=14401913

August 1985 OSHA Report - Carnation Company (1 Fatality)

EMPLOYEE #1 WAS CLEANING THE INTERIOR SURFACE OF A 8 FOOT BY 20 FOOT 68,000 POUND CAPACITY CORN OIL STORAGE TANK, WHICH PREVIOUSLY HAD CONTAINED CORN OIL COVERED BY A BLANKET OF NITROGEN GAS. EMPLOYEE #1 WAS WORKING

ALONE. HE WAS FOUND INSIDE THE TANK, UNCONSCIOUS, LYING FACE DOWN, BY A COWORKER.

https://www.osha.gov/ords/imis/accidentsearch.accident_detail?id=14382345

December 1985 OSHA Report - Ramco Plumbing Company (1 Fatality)

THE COMPANY HAD INSTALLED NEW DUCTWORK AND WAS REQUIRED TO PRESSURE TEST THE SYSTEM. THE EMPLOYEES FILLED THE SYSTEM WITH NITROGEN, BUT COULD NOT SUCCESSFULLY COMPLETE THE TEST. A HOLE WAS CUT IN THE DUCTWORK AND EMPLOYEE #1 CRAWLED IN TO LOOK FOR THE LEAK. HE LOOKED FOR THE LEAK IN THE 27 INCH DIAMETER DUCTWORK FOR ABOUT 15 MINUTES. AT ABOUT 10:30 AM EMPLOYEE #1 WAS DISCOVERED, DEAD, ABOUT 12 FEET FROM THE ENTRY HOLE.

https://www.osha.gov/ords/imis/accidentsearch.accident_detail?id=14400501

January 1986 OSHA Report Hydra Rig Cryogenics, Inc. (1 Fatality)

https://www.osha.gov/ords/imis/accidentsearch.accident_detail?id=14511547

Employee Entered Tank Filled With Nitrogen; Was Asphyxiated

April 1986 OSHA Report - Parker Industrial Services Inc / Exxon Baytown (1 Fatality)

Employee #1, age 19, was working atop a sulfur conversion unit, washing out the interior with a hose that was stuck down an open manway. The unit was under a nitrogen purge and Employee #1 was later discovered in the bottom of the unit. He died of asphyxia. Employee #1 was working along and without respiratory protection was used. There were no eyewitnesses to the accident.

https://www.osha.gov/ords/imis/accidentsearch.accident_detail?id=14498653

April 1986 OSHA Report - B. J. Transportation Service, Inc. (1 Fatality, 2 Nearly Overcome)

Employee #1 entered a semi tanker trailer to wash out the inside, which had contained a fatty acid (linseed oil). A short time later, he was found lying at the bottom under the hatch opening. Employees #2 and #3 jumped in to rescue him and were nearly overcome. Employee #1 died, and blood tests conducted two hours after exposure revealed carboxyhemoglobin levels of 56 percent for Employee #1, and 15 and 25 percent for Employees #2 and #3, respectively. The vessel's atmosphere was later found to be an inert gas containing mostly nitrogen and about 5 percent carbon monoxide. The gas had been made by the plant that unloaded the trailer, and had been used to blow off the load. When the driver arrived at the truck washing facility, he had not informed the workers about the inert gas, nor had the manifest indicated its use. The tank washing facility had not tested the atmosphere in the tank before Employee #1 entered, and there was no confined space entry program in effect.

https://www.osha.gov/ords/imis/accidentsearch.accident_detail?id=14385934

August 1986 OSHA Report Great Western Silicon Corp. (1 Fatality)

EMPLOYEE #1 WAS SANDBLASTING AT APPROXIMATELY 5:30 AM, USING A RESPIRATOR SANDBLASTING HOOD WHICH WAS HOOKED UP TO A COMPRESSD AIR LINE. THE LINE WAS SUPPLIED WITH NITROGEN. THE EMPLOYEE WAS ASPHYXIATED.

https://www.osha.gov/ords/imis/accidentsearch.accident_detail?id=979062

February 1987 OSHA Report - Himont U.S.A., Inc. (1 Fatality)

EMPLOYEE #1 WAS PLACING PHENOLIC NAME TAGS ON TESTING INSTRUMENTS INSIDE THE GAS CHROMATOGRAPH ROOM AT K-LINE. THE K-LINE GAS CHROMATOGRAPH WAS ON A NITROGEN BACK-UP FOR INSTRUMENT AIR FOR ITS ANALYZERS TO KEEP THEM EXPLOSION-PROOF. AT THE TIME, THE ANALYZERS HAD 10 SCFM OF POSITIVE PRESSURE FROM THE INSTRUMENTS BEING PURGED INTO THE GAS CHROMATOGRAPH ROOM. THE GAS CHROMATOGRAPH ROOM'S ATMOSPHERE THEREFORE WAS NITROGEN ENRICHED WHEN EMPLOYEE #1 ENTERED TO PLACE THE TAGS. THE NITROGEN REPLACED THE OXYGEN IN THE ROOM'S ATMOSPHERE, CAUSING AN OXYGEN DEFICIENCY. EMPLOYEE #1 DIED OF A LACK OF OXYGEN.

https://www.osha.gov/ords/imis/accidentsearch.accident_detail?id=14390983

June 1987 OSHA Report - Degussa Corporation (2 Fatalities)

Employee #1, a warehouse operator, dropped his safety glasses into the compartment of a railroad tank car that was being filled with aerosol. The aerosol was blown through a line by low pressure nitrogen gas. After Employee #1 notified coworkers and his supervisor, the material flow was turned off but not the gas. Wearing a gas mask, Employee #1 entered the railcar through a manhole. He was overcome and fell to the bottom of the tank. Employee #2, his supervisor, entered the railcar in a rescue attempt but was also overcome. Two other employees wearing gas masks attempted an unsuccessful rescue. Coworkers who donned SCBA removed the two bodies. Neither Employee #1 nor Employee #2 could be revived. Two hours later the tank air was tested and found to be at 14 percent oxygen.

https://www.osha.gov/ords/imis/accidentsearch.accident_detail?id=14353031

July 1987 OSHA Report - Kemanord, Inc. / Alabama Industrial Coatings, Inc. (2 Fatalities)

ON JULY 30, 1987, EMPLOYEES #1 AND #2 WERE SANDBLASTING IN A PIT UNDER A WEIGHING SCALE AT A CHEMICAL PLANT. THE EMPLOYEES WERE EMPLOYED BY A PAINTING CONTRACTOR. THE BLASTING EQUIPMENT WAS HOOKED TO A DIESEL-POWERED MOBILE AIR COMPRESSOR AND THEIR ABRASIVE BLASTING RESPIRATORS TO A COMPRESSED AIR LINE AT THE CHEMICAL PLANT. THE PLANT HAD LOST POWER AND ITS AIR COMPRESSORS SHUT OFF THE DAY OF THE ACCIDENT. THE COMPRESSED AIR LINES IN THE BUILDING HAD BEEN FILLED WITH PRESSURIZED NITROGEN. THEREFORE THE EMPLOYEES' ABRASIVE BLASTING RESPIRATORS WERE

SUPPLIED WITH NITROGEN INSTEAD OF AIR. THE EMPLOYEES DIED OF OXYGEN DEPRIVATION. THERE WAS ENOUGH OXYGEN IN THE PIT'S ATMOSPHERE TO SUPPORT LIFE, AS DEMONSTRATED WHEN SEVERAL EMPLOYEES SAFELY ENTERED THE PIT MINUTES AFTER THE ACCIDENT WITH NO RESPIRATORY PROTECTION.

https://www.osha.gov/ords/imis/accidentsearch.accident_detail?id=14528020

October 1987 OSHA Report - Big Three Industrial Gas, Inc. (1 Fatality)

When Employee #1 could not obtain breathing air from the installed line, he adapted unapproved hoses with quick disconnects so he could connect a respirator for use in abrasive blasting to a gas line supplying a blasting pot. This piping was not color coded or labeled in accordance with company policy, so Employee #1 did not know he was connecting to a nitrogen line instead of compressed air. He died of asphyxia. Since nitrogen is a waste gas in an air separation plant, it is used instead of compressed air for operating pneumatic equipment.

https://www.osha.gov/ords/imis/accidentsearch.accident_detail?id=14493290

January 1988 OSHA Report - Sta-Tek Corp. (1 Fataliy)

Employee #1 entered a bell jar to clean it wearing a Survivair Mark 2 airline respirator. The employer had decided to stop using removable, cleanable bell jar liners, and employees routinely wore supplied-air respirators when entering the bell jars to clean them with solvents. Management had anticipated a need for nitrogen gas in the airline piping and prepared a bypass to feed it in. Employee #1 was fed nitrogen gas through the completed bypass and died of asphyxia. No knowledgeable supervisor was available on site, and coworkers at the scene had not been trained in first aid or informed about the hazards of nitrogen gas. Key lock systems available at the bell jars were not used to bar entry.

https://www.osha.gov/ords/imis/accidentsearch.accident_detail?id=967729

March 1988 OSHA Report - Trinity Industries, Inc. - Plant #19 (East Plant) (2 Fatalities)

EMPLOYEES #1 AND #2 DIED OF ASPHYXIATION WHEN THEY ENTERED A RAILROAD TANK CAR THAT HAD BEEN PURGED WITH NITROGEN GAS. NONE OF THE OTHER EMPLOYEES KNEW WHY THE TWO WOULD HAVE ENTERED THE RAIL CAR, WHICH HAD BEEN LABELED "NITROGEN PURGE."

https://www.osha.gov/ords/imis/accidentsearch.accident_detail?id=14498240

March 1988 OSHA Report - Kerr-Mcgee Chemical Corporation/Pigment Plant (1 Fatality)

Employee #1 was to open a door to ventilate the room that contained the bag filter on top of a silo. The silo had been overfilled with titanium dioxide, which is transferred by a pneumatic conveying system that is operated at a high temperature (180 to 240 degrees Fahrenheit).

The overfilling flooded the room with hot titanium dioxide and air, causing several plastic lines containing nitrogen at 125 psi to rupture. The titanium oxide filled the room to a depth of 2 to 3 feet so that the door, which opens inward, had to be cut from its hinges to be opened. Once this was done (about three hours after the nitrogen lines had ruptured), Employee #1 pushed the door in at the top, creating a gap between the door and the door facing. He stepped through the gap into the room, presumably to shut off the nitrogen, but was overcome by the lack of oxygen. A helper with Employee #1 was not able to pull him out of the room alone. By the time help arrived, Employee #1 had died.

https://www.osha.gov/ords/imis/accidentsearch.accident_detail?id=14528244

May 1988 OSHA Report - Mr. Frank Inc. (1 Fatality)

EMPLOYEE #1 ENTERED A TANKER TRAILER, WHICH CONTAINED A RESIDUE OF SULFURIC AND NITRIC ACIDS, TO REPLACE A SAFETY VALVE. NE WAS NOT AWARE THAT COMPRESSED NITROGEN HAD BEEN USED IN THE UNLOADING PROCESS. HE WAS EQUIPPED WITH PROTECTIVE CLOTHING AND A FULL FACE ORGANIC VAPOR RESPIRATOR. WHEN EMPLOYEE #1 COLLAPSED UPON ENTERING, EMPLOYEE #2 WENT IN AFTER HIM AND ALSO COLLAPSED. EMPLOYEE #2 WAS QUICKLY PULLED OUT, BUT EMPLOYEE #1 COULD NOT BE REACHED FROM THE OUTSIDE. HE WAS NOT PULLED OUT FOR 30 MINUTES. THE RESPIRATORS WERE NOT ADEQUATE FOR AN OXYGEN DEFICIENT ATMOSPHERE. AIR TESTING AND VENTING WERE NOT DONE. PROBABLE CAUSES OF DEATH ARE AN OXYGEN DEFICIENCY OR THE PRESENCE OF NITROGEN DIOXIDE (CREATED BY THE BREAKDOWN OF NITRIC ACID AND DETECTED AT HIGH LEVELS).

https://www.osha.gov/ords/imis/accidentsearch.accident_detail?id=14377592

June 1988 OSHA Report - Industrial Corrosion Control (1 Fatality)

At approximately 4:00 p.m. on June 14, 1988, Employee #1, a contract painter, was in a JLG manlift (JBG0272) preparing to spray paint overhead pipe racks at a chemical plant. He was wearing a Bullard system 999 supplied air respirator, model 999-4 (long cape). Although he had been instructed not to use any of the plant's air lines for breathing air, he connected the respirator's air line to an outlet that was labeled "Tool Air--Do Not Breathe." The line was mislabeled and actually contained nitrogen. Employee #1 was found inside the manlift basket at ground level, dead of nitrogen asphyxiation.

https://www.osha.gov/ords/imis/accidentsearch.accident_detail?id=14528517

September 1988 OSHA Report - Boyd E. Hart Company, Incorporated (1 Fatality)

On September 8, 1988, Employee #1 and a coworker were sandblasting and painting grating and railings. Employee #1, the sandblaster, used an abrasive blasting respirator air line that was hooked into the plant's air supply. But the plant air supply was not grade D breathing air and was intended to be used only for valves, gauges, and pneumatic tools. Because the air compressor was shut down for maintenance, nitrogen backfed into the plant air line. No one from the company informed

Employee #1 or the painter that the line now contained nitrogen. Employee #1 donned the abrasive blasting helmet, inhaled the nitrogen, and was asphyxiated. He died.

https://www.osha.gov/ords/imis/accidentsearch.accident_detail?id=687632

September 1988 OSHA Report - Amoco Production Company (1 Fatality)

At approximately 8:00 a.m. on September 9, 1988, Employee #1, a part-time foreman, was sent to the top of a 146 foot high vessel to visually check the level of blown-in perlite insulation. A coworker, a laborer who had worked for the company for approximately 30 days, went with him. Two supervisors remained at ground level. Employee #1 and the coworker removed the manhole cover and were unable to see the perlite level. Employee #1 told the coworker not to tell anyone and then proceeded down into the vessel. The normal atmosphere had been displaced with 77 percent nitrogen. Employee #1 made it back to within four feet of the entrance point before he lost consciousness and fell back into the vessel. He was killed.

https://www.osha.gov/ords/imis/accidentsearch.accident_detail?id=733717

September 1988 OSHA Report - Boyd E. Hart Company, Incorporated (1 Fatality)

On September 8, 1988, Employee #1 and a coworker were sandblasting and painting grating and railings. Employee #1, the sandblaster, used an abrasive blasting respirator air line that was hooked into the plant's air supply. But the plant air supply was not grade D breathing air and was intended to be used only for valves, gauges, and pneumatic tools. Because the air compressor was shut down for maintenance, nitrogen backfed into the plant air line. No one from the company informed Employee #1 or the painter that the line now contained nitrogen. Employee #1 donned the abrasive blasting helmet, inhaled the nitrogen, and was asphyxiated. He died.

https://www.osha.gov/ords/imis/accidentsearch.accident_detail?id=687632

September 1988 OSHA Report - Baltimore Gas And Electric Company (1 Fatality, 1 Hospitalized)

Some electric utility employees were assigned to repair a level switch mechanism inside a 1136-kiloliter make-up water tank (condensate tank #11). The project required entry by a diver using a self-contained underwater breathing apparatus into the tank. Employee #2, the diver, obtained a diving permit and was evaluating means of entry as a predive drill to determine whether he could enter the manhole with his tanks and scuba or if they would have to be passed down to him after entry. He was wearing a lifeline when he went down the fixed ladder and into the tank. After he descended 203 millimeters, he was overcome and fell into the water. The diver tenders (Employee #1 and another employee) were holding the lifeline but could not see Employee #2 in the tank. Employee #1 entered the tank in an apparent rescue operation. He was not wearing a lifeline, nor was he using related emergency rescue equipment. The second diver tender, who was still holding the lifeline attached to Employee #2, yelled for assistance and, with the help of three coworkers, was able to raise Employee #2 to the bottom of the manhole opening. Since the line was around his lower chest, Employee #2's body was in a horizontal position, and the other employees had to reposition his body at an angle. The employees successfully completed the rescue and revived Employee #2 by administering cardio-pulmonary resuscitation. However, they could not see Employee #1 at this time. They summoned a professional diver, who found Employee #1 at the bottom of the tank. He had been asphyxiated. Employee #2 was hospitalized for his injuries. The dive team was not aware that a nitrogen blanket was in the space above the water. The team had not tested the atmosphere for oxygen. Employees were trained in confined space entry, but were not following entry procedures. In addition, the team had not obtained a confined-space permit, which would have required air monitoring from the Fire and Safety Department.

https://www.osha.gov/ords/imis/accidentsearch.accident_detail?id=889170

December 1988 OSHA Report - A & R Transport Inc. (2 Fatalities)

Two employees died from asphyxiation when they entered an oxygen deficient tank truck that was cleaned with nitrogen. The employees did not wear any respiratory protection or follow confined space procedures.

https://www.osha.gov/ords/imis/accidentsearch.accident_detail?id=14212971

October 1989 OSHA Report - Marathon Letourneau Company (1 Fatality)

On or about October 13, 1989, Employee #1, a melter B, entered a confined space to manually operate valves to a pressure vessel containing cal-sil and ferrites. Argon and nitrogen are used as a propellant for this mixture, and these lines were also located in the confined space. Employee #1 was overcome by the gases and died of asphyxia.

https://www.osha.gov/ords/imis/accidentsearch.accident_detail?id=767608

December 1989 OSHA Report - Austin Industrial, Inc (1 Fatality)

At approximately 10:30 a.m. on December 15, 1989, Employee #1 was preparing to replace the head of a vessel at approximately 40 feet above the ground when he saw a wrench lying on a plate 4 1/2 feet down inside the vessel. This was the vessel that he had removed the head from and had hooked up to a nitrogen purge line. Instead of moving the head in place with the cherry picker and bolting it down, he chose to bend over and enter the vessel to retrieve the wrench. As he started to pick up the wrench, he lost consciousness and collapsed on the plate where he was crouched. A second employee went for help because he did not have a vessel entry permit. Employee #1 was dead when he was finally retrieved.

https://www.osha.gov/ords/imis/accidentsearch.accident_detail?id=14219661

December 1989 OSHA Report - Custom Environmental Transport (1 Fatality, 1 Hospitalized)

Employee #1 was instructed to enter an LPG tank of a tractor-trailer truck. The tank had been purged with nitrogen approximate thirty minutes prior to Employee #1 entering it. Employee #1 was not warned of a hazardous situation, no atmosphere monitoring was done, and he was not equipped with respiratory protection equipment. He was overcome in the tank by inhaling residual nitrogen. The tank was probably oxygen deficient. Employee #2 entered the LPG tank to rescue Employee #1 and was asphyxiated. Employee #2, was apparently not trained in, or familiar with, the safe use of respirators, and entered the tank without the Scott Air-Pac 2 being properly worn. Employee #1 died and Employee #2 was hospitalized.

https://www.osha.gov/ords/imis/accidentsearch.accident_detail?id=14412308

December 1989 OSHA Report - Austin Industrial, Inc (1 Fatality)

At approximately 10:30 a.m. on December 15, 1989, Employee #1 was preparing to replace the head of a vessel at approximately 40 feet above the ground when he saw a wrench lying on a plate 4 1/2 feet down inside the vessel. This was the vessel that he had removed the head from and had hooked up to a nitrogen purge line. Instead of moving the head in place with the cherry picker and bolting it down, he chose to bend over and enter the vessel to retrieve the wrench. As he started to pick up the wrench, he lost consciousness and collapsed on the plate where he was crouched. A second employee went for help because he did not have a vessel entry permit. Employee #1 was dead when he was finally retrieved.

https://www.osha.gov/ords/imis/accidentsearch.accident_detail?id=14219661

January 1990 OSHA Report - Digital Equipment Corp. (1 Fatality)

Digital Equipment Corporation had established a central breathing-air station to supply air line respirators in three plant locations. An outside contractor was connecting a fourth location to the existing system using a nitrogen gas purge while braising copper line. Employees at one of the existing air line locations were not notified of the work going on. As a result, Employee #1 plugged in his respirator and was killed when he breathed nearly pure nitrogen.

https://www.osha.gov/ords/imis/accidentsearch.accident_detail?id=14269922

May 1990 OSHA Report - Sandy Hill South, Inc. (1 Fatality)

At approximately 5:30 p.m. on May 16, 1990, Employee #1 was using an airline respirator supplied by grade D breathing air from a compressed air cylinder while he prepared to spray paint. At approximately 3:10 p.m., Employee #1 switched to a full compressed-air cylinder. This change was witnessed by at least one other employee. Approximately 5 to 10 minutes later, Employee #1 was found unconscious. Two employees and paramedics performed CPR on the employee, but he died. The cylinder that Employee #1 was using was later analyzed and found to contain less than 1 percent oxygen and more than 98 percent nitrogen.

https://www.osha.gov/ords/imis/accidentsearch.accident_detail?id=14545487

July 1990 OSHA Report - Songer Corp (1 Fatality)

On July 15, 1990, Employee #1 and a crew, of Songer Corp., were welding inside a vessel boiler doing repair work. Air was being supplied by an air horn that was hooked up to the Great Lakes Steel air supply system. This system was used for cooling and ventilation. At 2:30 p.m., Employee #1 and the crew exited for a coffee break. At this time the air horn was removed from the manway opening and placed inside the vessel. Employee #1 returned to the vessel and, upon entering, was overcome by the lack of oxygen; nitrogen was induced into the compressed air line. Employee #1 died.

https://www.osha.gov/ords/imis/accidentsearch.accident_detail?id=170228431

July 1990 OSHA Report - Naval Air Station (1 Fatality)

Employee #1 was performing operational tests on the Mark-2 SCBA using a 9057-00 portable test console. The tests were being conducted with nitrogen gas rather than breathable air. Employee #1 was found dead with the face piece of the SCBA securely attached to his face and the SCBA regulator in the emergency position, allowing a constant flow of nitrogen gas into the face piece and his breathing zone. There were no witnesses. The sheriff's investigation is still open as of July 19, 1990; however, the sheriff's investigator stated that accidental death is indicated.

https://www.osha.gov/ords/imis/accidentsearch.accident_detail?id=740464

October 1990 OSHA Report - Midwest Elastomers (1 Fatality, 1 Serious Injury)

Employee #1 was found lying on his back, face up, in an open V-shaped hopper 6 ft by 6 ft by 6 ft. He either fell or climbed into the hopper and was overcome by nitrogen gas. Employee #1 died. Employee #2 was seen in the hopper yelling for help. It is believed that Employee #2 attempted to rescue Employee #1 and was himself overcome by nitrogen. Unfortunately, Employee #2 could not recall anything about the day of the accident and there were no witnesses. The hopper was filled with ground rubber pieces, which were to be run through a liquid nitrogen process that allows the rubber to be ground again. The hopper was nearly empty with an estimated 750 lb of rubber. It is believed that the nitrogen leaked back into the bin through the screw conveyor, which was attached to the bottom of the bin.

https://www.osha.gov/ords/imis/accidentsearch.accident_detail?id=14514616

April 1991 OSHA Report - Koch Refinery Rosemont, Minnesota (1 Fatality)

Employee #1 was assigned to clean the header bolts at the top of a reactor. The reactor was under a nitrogen purge. The vent port for the nitrogen was the header. A plastic cover was placed over the top of the reactor. The employee crawled under the plastic cover and died of asphyxia in the oxygen-deficient atmosphere.

https://www.osha.gov/ords/imis/establishment.inspection_detail?id=111683660

June 1991 OSHA Report - Chevron Usa Inc. (1 Fatality)

At approximately 2:00 p.m. on June 13, 1991, Employee #1 entered a 6 ft metal meter shed to calibrate a nitrogen flow recorder. The recorder is used to monitor the amount of nitrogen being injected into a gas well in order to enhance the recovery of gas and oil from underground formations. He was calibrating the flow recorder when, for some reason, a piece of stainless steel tubing came loose at a swage fitting, allowing nitrogen to rapidly enter the meter shed at approximately 4,800 psi. Employee #1 was quickly overcome in the nitrogen rich atmosphere and was killed.

https://www.osha.gov/ords/imis/accidentsearch.accident_detail?id=686261

July 1991 OSHA Report - Mobil Oil Corporation/Chalmette Refinery (1 Fatality)

At approximately 9:15 p.m. on July 22, 1991, the catalyst in the first stage hydrocracker reactor was being replaced, by a subcontractor, under a nitrogen atmosphere with 2 to 3 percent oxygen. No subcontractor employees were present at the time. Employee #1 came to the entry port of the unit without personal protective equipment, such as a respirator. Consequently, he was found unconscious on the top distributor tray and was transported to a local medical center. He never regained consciousness and died at approximately 4:30 p.m. on July 2, 1991. The probable cause of death was hypoxia.

https://www.osha.gov/ords/imis/accidentsearch.accident_detail?id=14414080

September 1991 OSHA Report - Gallo Winery (1 Fatality)

At approximately 8:30 a.m. on September 21, 1991, Employee #1, a maintenance painter, attached the breathing air hose for his supplied air respirator to a tee on a hose that had been previously connected to an unmarked compatible nitrogen quick-disconnect fitting. Coworkers found Employee #1 where he collapsed next to a filler unit in the bottling department. He died of asphyxia.

https://www.osha.gov/ords/imis/accidentsearch.accident_detail?id=511444

October 1991 OSHA Report - Superior Graphite Co. (2 Overcome by Nitrogen – both rescued) (2 Rescued by local fire dept).

Employee #1 entered a 9 ft deep furnace that measured between 42 in. and 33 in. in diameter to retrieve a broken electrode. Employee #1 was overcome by nitrogen gas. Employee #2 attempted to rescue him and also fell unconscious. Both men were rescued by the local fire department.

https://www.osha.gov/ords/imis/accidentsearch.accident_detail?id=782672

December 1991 OSHA Report - Bob Eisel Industrial & Commercial Coating, Inc. (1 Fatality)

Employee #1 hooked the fresh air line of his air supplied respirator into a plant's compressed air lines, then began some abrasive blasting. The plant operators, unaware that their plant air was being used as breathing air, shut down the fresh air compressor for routine, scheduled maintenance and pumped nitrogen into the system to maintain pressure and control the valves in the refinery. The employee was overcome by the nitrogen in the air lines and died of nitrogen asphyxia.

https://www.osha.gov/ords/imis/accidentsearch.accident_detail?id=14422943

April 1992 OSHA Report - Alumax Mill Products (3 Fatalities)

In April 1992, Employee #1, of CTC, was overcome when he entered a coating tank to clean it. Employees #2 and #3, of Alumax, were also overcome when they entered the tank to attempt a rescue. The tank had been tested and ventilated on the day before the incident but not tested on the day of it. On the day of the incident, the tank had been ventilated with what was thought to be compressed air but was actually nitrogen. All three employees died of asphyxiation.

https://www.osha.gov/ords/imis/accidentsearch.accident_detail?id=767681

May 1992 OSHA Report - Arcadian Corporation (1 Fatality)

Employee #1 and two coworkers were cleaning filters in a hydrogen purifying tank. Employee #1 used a self-propelled personal lift to access the tank's upper external portion. While on the lift, he inserted his upper body into the tank. During the cleaning operation, the tank was partly purged with nitrogen to remove dust particles from inside the tank's upper portion. During that time, the coworkers noted that the employee was not responding. Using a crane to access the tank's external upper level, the coworkers found Employee #1 unconscious. He later died of oxygen deficiency.

https://www.osha.gov/ords/imis/accidentsearch.accident_detail?id=829234

June 1992 OSHA Report - Union Drilling Company (2 Fatalities)

On the morning of June 3, 1992, the bodies of Employees #1 and #2 were found by coworkers inside a tank truck used to recover gas well fracing chemicals. The cause of death was listed as asphyxiation due to exposure to acid fumes and oxygen deprivation. At the time of the accident, 1 1/2 million cubic feet of nitrogen gas, which was used in the fracing process, was being piped from the wellhead to the tank to recover fracing chemicals, including hydrochloric and acidic acids. Willful citations were issued to two employers for confined space entry and hazard communication deficiencies, along with other alleged violations. A third employer was issued a citation for hazard communication deficiencies under the multi-employer workplace provision in the standard.

https://www.osha.gov/ords/imis/accidentsearch.accident_detail?id=14322119

November 1992 OSHA Report - Sparta Fruit Storage Inc (1 Fatality)

Employee #1 and two coworkers were working at an apple storage facility. At 11:30 a.m., one of the coworkers left for the day to deliver cider. At 2:00 p.m., the other coworker's day ended and she went home. At this point only Employee #1 remained at the plant and was assigned to clean up broken wooden crates. At approximately 6:35 p.m. the coworker delivering cider returned to the facility and noticed that the lights and a radio were on in the building. As he went to turn off the lights and radio, he noticed that the 2 ft by 2 ft plexiglass window installed in apple storage room #4 was missing. He found Employee #1 unconscious inside the room. The room is a controlled atmosphere room where apples are stored in an atmosphere of less than 3 percent oxygen, 3 percent carbon dioxide, and more than 94 percent nitrogen. Employee #1 died due to exposure to the oxygen-deficient atmosphere. He was not to enter the room. It is not known why he entered the room.

https://www.osha.gov/ords/imis/accidentsearch.accident_detail?id=912576

August 1993 OSHA Report - Lion Oil Company (1 Fatality)

Employee #1 was asked by an inspector to go up on the platform below the #9 reactor and remove some catalyst that had been spilled during the catalyst removal process. Pipefitters had worked from this platform during the removal of eight catalyst downspout pipings. The pipefitters had worn fresh air respirators during that process because of the nitrogen purge placed in the reactor. Once all of the downspouts had been removed and blind flanges installed, the pipefitters had removed their fresh air masks and worked another 15 or 20 minutes. Employee #1, who was working without a respirator, was found unconscious after working only a few minutes on this platform. He was later pronounced dead. Autopsy results are still pending.

https://www.osha.gov/ords/imis/accidentsearch.accident_detail?id=564906

October 1993 OSHA Report - Metro Electrical Contractors Inc. (1 Fatality, Several others exposed)

Employees #1 and #2 were assigned to adjust the setting on a level data pressure sensor that was mounted on top of a soybean storage tank in order to regulate the low pressure nitrogen used to blanket the tank. The storage tank was 24 ft high by 30 ft diameter with a 262,000 gallon capacity. There was 1 in. of oil in the tank. Employee #2 had removed trash from the area and when he returned he could not locate Employee #1. After a short time Employee #2 climbed the tank ladder and, when he reached the top, he could see that the tank hatch was open. He looked in the tank and saw Employee #1 lying in the bottom. He yelled to Employee #3 for help, then entered the tank to assist Employee #1. Employee #3 climbed the tank and could see both Employees #1 and #2 lying in the bottom. He yelled for help, shut off the nitrogen flow to the tank, and climbed down. He was removing a side plate on the tank when coworkers arrived to assist. After removing the plate, Employee #3 held his breath and entered the tank and pulled Employee #2 to the opening. He was able to get out of the tank before being completely overcome. Employee #4 entered the tank and was able to pull Employee #1 out of the tank before being was completely overcome by the nitrogen atmosphere. Employee #1 died of asphyxia.

https://www.osha.gov/ords/imis/accidentsearch.accident_detail?id=14495154

November 1993 OSHA Report - 3-R Services, Inc. (1 Fatality)

At approximately 10:45 p.m. on November 5, 1993, Employee #1 was on top of a frac tank that was being used to recover the backflow from an oil well. Employee #1 was measuring the fluid level of the frac tank using a tank gauge. The backflow from the oil well consisted of water, compressed nitrogen gas, and a little sand. Employee #1 was using a flashlight to peer into an opening that was approximately 2 ft by 2 ft. A coworker was assisting Employee #1 by opening the flow line at the oil well. Employee #1 apparently dropped the flashlight into the opening and climbed into the frac tank to retrieve it. The coworker, who was approximately 65 ft away, went over to the tank to see why Employee #1 was not on top of the tank. Employee #1 had been overcome by the gas in the frac tank. The coworker attempted to rescue Employee #1 from the frac tank but was not successful. Employee #1 died.

https://www.osha.gov/ords/imis/accidentsearch.accident_detail?id=566612

January 1994 OSHA Report - Quala Systems, Inc Dba Quala Wash (1 Serious Injury)

At approximately 4:15 p.m. on January 21, 1994, Employee #1 engaged in cleaning a single compartment 5,000 gallon tank trailer with a residue of dodecylbenzenesulfonic acid that had been purged with nitrogen. The tank was at Bay 2. He entered the tank in his wash suit, boots, harness, and full face dual cartridge acid gas/organic vapor respirator, and holding a garden hose. He had been trained in company policy that all dirty tanks were permit-required confined spaces. He acted contrary to company policy and entered without a permit, attendant, oxygen meter, winch retriever, or lifeline. Suffering asphyxiation, Employee #1 required hospitalization. He acted independently of company expectations. No T8CCR violations were attributed to the accident.

https://www.osha.gov/ords/imis/accidentsearch.accident_detail?id=170374607

Apil 1994 OSHA Report - G. Heileman Brewing Co (1 Fatality)

Employee #1was assigned to clean a nitrogen-inerted 900 gallon tea tank. There was not sufficient oxygen present and Employee #1 was overcome and killed when he entered the tank.

https://www.osha.gov/ords/imis/accidentsearch.accident_detail?id=14372759

September 1994 OSHA Report - Freeze Systems of Delaware, Inc. (1 Fatality)

Employee #1 was checking the temperature of a pipe that was being frozen with liquid nitrogen. The employee either fell or went into the trench where the pipe was located. The employee was exposed to a nitrogen-enriched/oxygen-deficient atmosphere and was asphyxiated. Employee #1 died.

https://www.osha.gov/ords/imis/accidentsearch.accident_detail?id=820407

October 1994 OSHA Report - Catalyst Services, Inc. (1 Fatality)

Employee #1 was removing catalyst from a reactor at the Conoco Petroleum Refinery. The reactor was overpressurized with nitrogen and when he broke through the crust layer at the top of the catalyst, the pressure release forced the employee out of the reactor. He suffered multiple trauma injuries and died.

https://www.osha.gov/ords/imis/accidentsearch.accident_detail?id=882670

January 1995 OSHA Report - Dave Evans Welding (2 Fatalities)

Employee #1 was watching a Crysen employee retrieve a catalyst sample from the reactor of a MDDW unit that had been purged with nitrogen. Employee #1 had designed and made the sample tool the Crysen employee was using and was there to make sure that it worked as it should. As the Crysen employee came out of the reactor so that workers could change a breathing air cylinder, he either accidentally disconnected the air line to his breathing apparatus or ran out of air. He stood up to grab the chain ladder, then fell over and appeared to be unconscious. The Crysen maintenance supervisor ran to an air bottle that was about 15 feet away to check the air, which was already dangerously low. When the supervisor got back to the vessel opening, Employee #1, who was not wearing a breathing apparatus, had already jumped into the vessel. The maintenance supervisor and the engineer consultant, who was also at the scene watching the sampling, went to the opening, saw Employee #1 attempting to lift the Crysen employee to the opening, and were able to grab the Crysen employee's arm. At that time, the engineer consultant saw Employee #1 go down. They attempted to pull the Crysen employee out, but he seemed to be hung up on something. The nitrogen coming out of the hole was so concentrated that the supervisor and consultant started to pass out, and could not hold on any longer, and let go of the Crysen employee. Once they were away from the opening, the maintenance supervisor radioed for help and three employees from the control room came to assist. One of them attempted to enter the vessel with a SCBA, but the opening was only 18 in. wide, and he could not get through. Rescue attempts were stopped until the fire department arrived. However, the ambulance report stated that Employee #1, and the Crysen employee, both of whom had died, were out of the vessel when it arrived.

https://www.osha.gov/ords/imis/accidentsearch.accident_detail?id=170681746

March 1995 OSHA Report - Ciba Pharmaceuticals (1 Fatality)

At approximately 12:55 p.m. on March 15, 1995, Employee #1, a chemical operator at Ciba-Geigy in Summit, NJ, was found slumped in the manway of reactor XR30. According to the batch sheet, the employee had been dry charging bromoketone powder into the nitrogen-inerted reactor. The medical examiner determined that the employee died of cerebral anoxia due to inhaling nitrogen gas.

https://www.osha.gov/ords/imis/accidentsearch.accident_detail?id=170022818

June 1995 OSHA Report - Rescar, Incorporated (1 Fatality, 2 Hospitalized)

Employee #2 was assigned to inspect the interior of a new railroad tank car that had just arrived at the facility the night before. Employee #2 and his attendant followed the company's confined space program by getting a permit and oxygen/LEL meter. They put the meter on a broom stick and waved it around inside the tank, listening for an audible alarm. They then brought the meter out into the outside air and took the oxygen reading, which was 20.8 percent. Employee #2 then entered the tank car with no respiratory protection. He was unconscious within a minute. The attendant called over Employee #1, who also went in and passed out. Employee #3 tried to enter, but became dizzy and had to be helped out. The local fire department arrived with SCBAs and rescued Employees #1 and #2. Employee #1 died and Employees #2 and #3 were hospitalized. The company later learned that the car contained a nitrogen blanket to inert the atmosphere.

https://www.osha.gov/ords/imis/accidentsearch.accident_detail?id=552430

June 1995 OSHA Report - Guilford Mills, Inc. (1 Fatality)

At approximately 9:28 a.m. on June 12, 1995, Employee #1, age 23, was found trapped up to his waist inside a permit-required confined space. He was unconscious with no vital signs. CPR was administered but Employee #1 was pronounced dead at the scene. The atmosphere within the confined space was suspected to be contaminated with nitrogen because of a bypass valve that was continually putting nitrogen into the confined space at 2 psi. Two hours after the body was removed, an atmospheric measurement of the confined space showed a 6 percent oxygen level.

https://www.osha.gov/ords/imis/accidentsearch.accident_detail?id=170060917

August 1995 OSHA Report - Sentell Bros Inc (1 Fatality)

Employee #1 connected a supplied air breathing hood line to a plant nitrogen line and then donned the hood to start sandblasting. He passed out and was noticed by a coworker about two minutes later. CPR was administered to Employee #1 but was unsuccessful. He died of asphyxiation.

https://www.osha.gov/ords/imis/accidentsearch.accident_detail?id=978296

October 1995 OSHA Report - Minnesota Rubber Company, Div. Of Quadion Corp. (1 Fatality)

Employee #1 entered a confined space where liquid nitrogen was being used in a cooling process. He was not wearing respiratory protection and was overcome by liquid nitrogen and oxygen deficiency. He died of asphyxiation. Temperature in the space was approximately negative 3 degrees Fahrenheit to negative 5 degrees Fahrenheit and oxygen content was approximately 4.9 percent to 6 percent.

https://www.osha.gov/ords/imis/accidentsearch.accident_detail?id=906578

March 1996 OSHA Report - Morris Bean & Company (1 Fatality)

Employee #1 was using an air hammer to chip residues out of furnace #3 at an aluminum foundry. He was wearing a Bullard (#88 series) air line respirator. Two compressed gas lines with universal access couplings were attached to post #1-5. The one on the right was labeled "natural gas". The one on the left had an old paper tag attached, with the word "air" handwritten on it; however, this line actually contained pure nitrogen. A splitter diverted one part of the gas stream to the air hammer, and the other part of the stream to the air line respirator. Employee #1 was asphyxiated and killed when exposed to pure nitrogen.

https://www.osha.gov/ords/imis/accidentsearch.accident_detail?id=14370985

July 1996 OSHA Report - Bronco Winery (1 Serious Injury)

Employee #1, age 20, inadvertently connected the air hose of the hood he was wearing to a nitrogen line because it was equipped with fittings compatible with his air line. He suffered from oxygen deficiency and asphyxia.

https://www.osha.gov/ords/imis/accidentsearch.accident_detail?id=170762207

October 1996 OSHA Report - Transport Service Co. (2 Fatalities)

On October 25, 1996, Employee #1 entered a 6,340 gal tank-truck trailer to visually inspect the tank's interior from the hatchway. He broke the entry plane, lost consciousness, and fell in. This was witnessed by Employee #2, who entered the tank to rescue Employee #1. Employee #2 also lost consciousness and fell in. Employees #1 and #2 died from chemical asphyxiation. The tank was filled with nitrogen gas, creating an oxygen-deficient atmosphere.

https://www.osha.gov/ords/imis/accidentsearch.accident_detail?id=14408835

October 1996 OSHA Report - Philip Services Corporation (1 Fatality)

At approximately 11:30 a.m. on October 30, 1996, Employee #1, a tank washer, entered a truck tanker trailer to spot treat the inside of the tank with acid cleaner. He was found unconscious in the tank 10 minutes later, and was pronounced dead at the scene. The tank had contained a shipment of mercaptobenzothiazole that had been off-loaded using nitrogen as a pad. The tank had then been mechanically cleaned. Employee #1 was not wearing a respirator, and it was unclear if he monitored the atmosphere before he entered.

https://www.osha.gov/ords/imis/accidentsearch.accident_detail?id=170570402

December 1996 OSHA Report - Taurus Service, Inc. (2 Fatalities)

At 7:00 a.m. on December 6, 1996, Employees #1 and #2, who had been transporting frozen bull semen in a white commercial van, were found unconscious by employees of Longway Restaurant. New York State police troopers responded to the scene and determined they were dead. Subsequent investigation found that there was a 100 gallon container of liquid nitrogen inside the van that had vented and/or leaked to the interior, displacing the oxygen.

https://www.osha.gov/ords/imis/accidentsearch.accident_detail?id=200880052

February 1997 OSHA Report - Inchem Corporation (2 Fatalities)

Employees #1 and #2, ages 21 and 19 years, respectively, were working a weekend night shift without onsite supervision. They were told at the beginning of their shift to prep a tanker trailer that was to be loaded with product the next day. They were instructed not to go into the tanker, but to use rags attached to a long stick to wipe up some water from the bottom. The next morning, Employees #1 and #2 were found dead inside the tanker. They died after inhaling nitrogen from the nitrogen supply line that was attached to the feed line into the tanker. The nitrogen gas was flowing when the employees were found.

https://www.osha.gov/ords/imis/accidentsearch.accident_detail?id=170570782

November 1997 OSHA Report - Hill Air Force Base, D.O.D (1 Fatality)

Employee #1 was asphyxiated while working in a spray booth. Employee #1 attached his supplied air respirator to a pure nitrogen supply. The nitrogen supply in the booth was configured with breathing air female fittings compatible with male fittings used with the supplied air respirator hoses. Due to this configuration, Employee #1 was able to insert the respirator coupling into the nitrogen supply. After breathing pure nitrogen for an unknown period of time, Employee #1 lost consciousness and died.

https://www.osha.gov/ords/imis/accidentsearch.accident_detail?id=201570173

March 1998: CSB Reported Union Carbide Fatality (1 Fatality, 1 Injury)

https://www.csb.gov/union-carbide-corp-nitrogen-asphyxiation-incident/

Employees #1 and #2 were part of a crew installing an oxygen feed mixer on the fifth floor of the etox unit. An enclosure had been made by wrapping the opening of a large pipe with black plastic, and Employees #1 and #2 entered this enclosure to check with a black light the surface of the pipe's flange for oil contamination. The pipe was filled with nitrogen, which was being put into the process equipment to prevent rusting. Employees #1 and #2 were overcome in the oxygen-deficient atmosphere. Coworkers in the immediate area could not see through the black plastic that the men were in trouble, and so did not respond in a timely manner. Finally, a contractor determined that there was a problem and the plant's emergency response team arrived and provided emergency medical treatment. Employees #1 and #2 were transported to the hospital, where Employee #1 was pronounced dead. Employee #2 was still recovering from the incident and appeared to be suffering from short-term memory loss.

https://www.osha.gov/ords/imis/accidentsearch.accident_detail?id=170080444

December 1998 OSHA Report - Refractory Maintenance Corporation (1 Fatality)

Employee #1 was found, unconscious and not breathing, inside the cooling chamber of a brazing oven. He was transported to the hospital, where he died. He had climbed into the chamber to retrieve a cable that was attached to the conveyor belt. A nitrogen gas valve for the oven was found to be cracked open, suggesting that Employee #1 was overcome by the nitrogen-rich environment.

https://www.osha.gov/ords/imis/accidentsearch.accident_detail?id=200880342

April 1999 OSHA Report - J.A.Z. Inspection Corp. (1 Fatality)

Employee #1 was testing lauric acid on a vessel when his sampling device fell into the tank. He entered the vessel to retrieve it and was overcome by nitrogen gas, which was being used as a blanket. Employee #1 dies of asphyxia. He was mot using any form of personal protection equipment when he entered the tank.

https://www.osha.gov/ords/imis/accidentsearch.accident_detail?id=200021228

June 1999 OSHA Report - National Service Company of Iowa (1 Fatality)

Employee was in the meat processing room when nitrogen was released from an unlabelled pipe. He died of asphyxia.

https://www.osha.gov/ords/imis/establishment.inspection_detail?id=302342407

June 1999 OSHA Report - Boc Gases (1 Fatality)

Employee #1 was working in the small Urich building where nitrogen and other substances are used to fill compressed cylinders by a pressurized hosing systems connected to outside bulk storage tanks. Apparently when Employee #1 came to work and was alone, he found that the nitrogen hose had not been shut off the night before and had leaked out a large quantity of nitrogen. Employee #1 attempted to shut off this valve to the hose and was asphyxiated due to the displacement of oxygen in this room by the nitrogen leak. The shut-off valve for this bulk tank was also found partially open.

https://www.osha.gov/ords/imis/accidentsearch.accident_detail?id=201953171

May 2000 OSHA Report - Six Ks Painting Contractors Inc. (Hospitalized Injury)

Employee #1 plugged his supplied-air respirator into a nitrogen source. He was hospitalized for treatment of asphyxia.

https://www.osha.gov/ords/imis/accidentsearch.accident_detail?id=200021830

August 2000 OSHA Report - Defiance Testing & Engineering Services Inc (1 Fatality)

Early in the morning of August 15, 2000, an employee was found frozen to death at his workplace. The employee had reportedly entered a nitrogen chamber used to perform tests on automotive components. The chamber was being maintained at temperatures below -35 degrees Fahrenheit for the purpose of performing the tests. The medical examiner determined the cause of death to be hypothermia; however, the possibility of asphyxia or cardiac arrest due to nitrogen toxicity could not totally be excluded.

https://www.osha.gov/ords/imis/accidentsearch.accident_detail?id=201640042

September 2000 OSHA Report - Ge Medical Systems (1 Fatality)

Employee #1 was working alone to install an MRI. The process required cryogenic nitrogen to cool the magnet. While the nitrogen was being applied to the magnet, it apparently leaked into the work area. Employee #1 died of asphyxia.

https://www.osha.gov/ords/imis/accidentsearch.accident_detail?id=200851624

December 2000 OSHA Report - Carriage-By-The-Lake Of Ihs, Inc. (4 Fatalities – all nursing home patients)

On December 7, 2000, the Cincinnati OSHA Office heard through the media and through police reports that two fatalities had occurred at a nursing home in Bellbrook, Ohio. An OSHA compliance officer was sent up to the site to begin to determine what had gone wrong, whether or not there was occupational exposure to a hazard, and if OSHA should play a lead, supporting or no role in investigating this tragedy. The conclusion was reached that OSHA should not play a lead role in this investigation because the exposures and the deaths occurred only to patients, not to employees, of the nursing home. This determination was made as FDA was actively investigating the incident and taking a lead role in performing an investigation. The nursing home routinely ordered and received large compressed gas cylinders, or tanks, containing pure oxygen, for consumption by some of their residents, such as emphysema patients, who have unhealthy respiratory systems. Their supplier, BOC Gases, mistakenly delivered one tank of pure nitrogen in addition to the three tanks of pure oxygen which had been ordered. An employee at the nursing home hooked up this tank, which contained pure nitrogen, to the nursing home's oxygen delivery system. On December 7, 2000, this event caused two nursing home residents to die, and three additional nursing home residents were admitted to hospitals in critical condition. Within the following month, two of these three additional residents also died, bringing the total death toll to four.

https://www.osha.gov/ords/imis/accidentsearch.accident_detail?id=837914

March 2001 OSHA Report - Quala Systems, Inc. (1 Fatality)

Employee #1 entered a permit-required confined space (tanker truck) without following confined space procedures. Nitrogen had been used in the tanker to help unload the product. The nitrogen displaced the oxygen in the tanker. Employee #1 entered into an oxygen-deficient atmosphere and died of asphyxia.

https://www.osha.gov/ords/imis/accidentsearch.accident_detail?id=200101103

March 2001 OSHA Report - Magnex Scientific Limited (1 Fatality)

On September 20, 2001, Employee #1 was installing an MRI machine. During installation Employee #1 was using liquid nitrogen to cool the magnet. The nitrogen was not properly ventilated and displaced the oxygen in the room. Employee #1 died of asphyxiation.

https://www.osha.gov/ords/imis/accidentsearch.accident_detail?id=202340022

April 2001 OSHA Report - E I Dupont De Nemours & Company (1 Fatality)

Employee #1 and another employee were preparing a batch load of dry chemical powders and crystals for mixing and transfer in the two level engineering polymers mixing facility. One employee left the upper level and began preparing the lower level additives for the mix. When the control room operator noticed that no weight was being added, the lower level employee was instructed to check on Employee #1. The lower level employee found Employee #1 down on the floor with a nitrogen gas hose attached to his respirator hood hose. Employee #1 died from asphyxiation.

https://www.osha.gov/ords/imis/accidentsearch.accident_detail?id=200621589

July 2001 OSHA Report - Precision Mechanical, Inc. (1 Fatality)

On July 8, 2001, Employee #1, a pipefitter, was disassembling a section of non-flanged 6-in.-diameter light-weight piping. The piping was under the pressure of nitrogen gas served to provide a positive pressure purge on an adjacent recently chemically-cleaned section of permanent piping. The section of pipe being disassembled had not been isolated from the source of the nitrogen gas. No lockout-tagout system was in place to assure that gas was not diverted into the loop being disassembled. Additionally, the unrestrained section of light-weight pipe was adjoined to a twisted section of flexible hose which likely placed mechanical pressure on the pipe. The pipe blew up in the pipefitter's face when he removed a bolt from the clamp holding a coupling device in place. He suffered fatal injuries.

https://www.osha.gov/ords/imis/accidentsearch.accident_detail?id=102329653

August 2001 OSHA Report - R.S.I. Inc. (1 Fatality)

At approximately 2 a.m. on August 8, 2001, Employee #1 was working on top of a reactor under nitrogen purge. A coworker was wearing breathing air with a communication device. The lead man was on top of the reactor overseeing the job. The lead man turned around and began communicating on the headset with the workers monitoring the breathing air. Employee #1 walked past the lead man and the coworker without breathing air and reached into a manhole. Employee #1 was overcome by the fumes/vapors and fell 3 to 4 feet into the reactor. The coworker and lead man pulled Employee #1 out of the reactor, and the lead man performed CPR until the rescue team arrived. Employee #1 was pronounced dead at the hospital. Employee #1 died from asphyxiation.

https://www.osha.gov/ords/imis/accidentsearch.accident_detail?id=123276610

March 2002 OSHA Report - George Young & Sons, Inc (Serious Injuries)

On March 5, 2002, Employees #1 and #2 with George Young & Sons were preparing to perform abrasive blasting work in a second floor spray chamber at Keystone Steel and Wire. They were instructed by the site foreman to don their type CE abrasive blast respirators and upon his return to the area, he found Employees #1 and #2 unconscious. The supplied air line for the respirators had been connected to a plant air line labeled "compressed air". After investigating, it was found that the line was 99.998 percent nitrogen. Both employees were hospitalized.

https://www.osha.gov/ords/imis/accidentsearch.accident_detail?id=200271013

May 2002 OSHA Report - Bently Nevada, LLC. (1 Fatality)

On May 22, 2002, Employee #1, a machinery maintenance man, entered a shed that was being purged of nitrogen gas. The shed consisted of metal instrumentation and measured approximately 8-feet long by 8-feet wide by 10-feet tall. A short time later, Employee #1 was discovered unconscious, lying on the shed floor. He was pronounced dead due to asphyxiation.

https://www.osha.gov/ords/imis/accidentsearch.accident_detail?id=201771706

November 2002 OSHA Report - Hb Fuller Co (1 Fatality)

On November 8, 2002, Employee #1 was found with his upper torso inside a testing hatch on a reaction vessel where a polyol was being manufactured. The apparent cause of death was asphyxiation from the nitrogen purge in the reaction vessel.

https://www.osha.gov/ords/imis/accidentsearch.accident_detail?id=201640059

July 2004 OSHA Report - Hospira, Inc. (1 Fatality)

On July 1, 2004, Employee #1, a lyophilizer (freeze dryer) operator, entered the chamber of the machine to perform lubrication, which was required under the preventative maintenance program. Employee #1 was not authorized to perform lockout procedures and so the nitrogen was not isolated. When co-workers found Employee #1 in a non-responsive state they noticed that nitrogen was being released into the confined space where Employee #1 was working.

https://www.osha.gov/ords/imis/accidentsearch.accident_detail?id=201261641

November 2004 OSHA Report - Quality Carriers Inc (1 Fatality)

At approximately 4:00 p.m. on November 30, 2004, Employee #1 was working in a tanker trailer that was located in the second bay of a shop to repair a leaking valve. The shop foreman discovered Employee #1 inside the tanker unconscious. Emergency services were contacted, and Employee #1 was transported to a local hospital, where he was declared dead due to asphyxiation. The accident investigation revealed that the air was not tested prior to Employee #1 entering the tanker. It was determined that the tank had contained an unknown concentration of nitrogen, therefore reducing the safe level of oxygen in the space.

https://www.osha.gov/ords/imis/accidentsearch.accident_detail?id=201390994

November 2005 OSHA Report - Matrix Service Industrial Contractors, Inc. / Valero Delaware City Refinery (2 Fatalities)
(See the CSB report on this incident)

Employees #1 and #2, two boilermakers, were working on top of a reactor which was being overhauled. The reactor was inerted with nitrogen. The employees were removing a roll of tape which had been left in the reactor. It is presumed that Employee #1 either fell or climbed into the reactor and then Employee #2 climbed in to rescue Employee #1. Both employees were killed.

https://www.osha.gov/ords/imis/accidentsearch.accident_detail?id=200411049

January 2007 OSHA Report - Swarthmore College (1 Fatality)

Between 9:30 a.m. and 1:00 p.m. on January 19, 2007, Employee #1 inserted his upper torso into a cardboard box. He was probably adjusting the placement of plastic tubing of which one end was attached to a wall mounted needle valve that was dispensing nitrogen. The other end was inserted into the box through a cut out opening. He was alone in a room with a self locking door. At approximately 8:30 p.m. a security officer found him in a non-responsive condition. Employee #1 died of asphyxiation due to the displacement of oxygen by nitrogen.

https://www.osha.gov/ords/imis/accidentsearch.accident_detail?id=202004891

April 2007 OSHA Report - Nucor Steel (1 Fatality)

On April 5, 2007, a Melt Shop Mechanical Maintenance employee was working alone replacing three filters in the baghouse on top of the Asbury for the Furnace Number 1. The employee had not made radio contact for several hours so the Supervisor went to the area where he was working. The employee was found in the top of the baghouse crouched over in a nitrogen rich atmosphere. The employee was pronounced dead at the scene and was estimated to have been dead for about 2 hours.

https://www.osha.gov/ords/imis/accidentsearch.accident_detail?id=200263184

May 2007 OSHA Report - Clinton Safety Associates, Inc. (1 Fatality)

On May 12, 2007, an employee entered into tanker trailer Number 1208 at the CPJ Technologies facility in order to clean out sludge that had accumulated in the bottom of the tanker. The tanker had been inerted with nitrogen, thus creating an IDLH atmosphere. At approximately 11:45 am, the employee lost air flow to his supplied air hood. He yelled to the attendant at the top of the tanker that he had no air, but there was only a manual retrieval system available and the attendant could not extract the employee from the confined space. The emergency medical services were called, and the Taylors Fire Department arrived at approximately 11:50 a.m. They removed the employee from the tanker and found him dead.

https://www.osha.gov/ords/imis/accidentsearch.accident_detail?id=200374221

August 2007 OSHA Report - Quala Systems, Inc. Dba Qualawash (1 Fatality)

On August 4, 2007, Employee #1, a tank cleaner, entered an oxygen deficient atmosphere inside a 6,340 gallon intermodal (ISO) container, which had previously contained ethyl acetate, and had been unloaded using nitrogen. Employee #1 wore an air-purifying respirator with multi-gas/vapor/P100 cartridges and died of asphyxia.

https://www.osha.gov/ords/imis/accidentsearch.accident_detail?id=200902070

October 2007 OSHA Report - Arkema Inc. (1 Fatality)

On October 30, 2007, Employee #1, an electronic technician, was asphyxiated by a nitrogen line leak in the analyzer building.

https://www.osha.gov/ords/imis/accidentsearch.accident_detail?id=201855988

June 2008 OSHA Report - Blommer Chocolate Company (1 Fatality, 2 Injured)

At approximately 7:00 a.m. on June 8, 2008, Employee #1 started a cooling process for a ribbon blender containing black cocoa powder. The cooling process utilized large quantities of compressed liquid nitrogen to lower the temperature of the black cocoa powder. The ribbon blender had a ventilation system designed to remove excess nitrogen from the blender safely. However, on that particular day, the ventilation system became compromised. The nitrogen, which entered into the ribbon blender, exceeded the blender capacity and backed up into an attached, unsealed screw conveyor. The nitrogen leaked out of the screw conveyor and into the black cocoa control room, causing an oxygen deficient atmosphere. At approximately 9:00 a.m. on June 8, 2009, Employee #1 entered the black cocoa control room. A surveillance video of the area showed that within minutes of entering the control room, Employee #1 became unconscious. At approximately 10:00 a.m. on June 8, 2008, Employee #2 entered the control room and quickly became unconscious. Employee #2 remained unconscious for approximately one hour after entering the control room. For unknown reasons, Employee #2 became semiconscious and used a phone to report Employee #1 as deceased and report their location. A supervisor quickly called 911, and employees quickly responded to the black cocoa control room. The employees who responded were unaware of the asphyxiation hazard and briefly entered the black cocoa control room. The responding employees told Employee #2 to remain in the control room, while they left to meet the outside emergency responders. Meanwhile, Employee #2 left the control room. Employee #2 was eventually located in a nearby stairwell, semiconscious. Employee #3, who was one of the first employees to reach the accident scene, became winded from running up and down stairs and was prompted by the Fire Department to seek medical attention at the hospital. Employee #1 was pronounced dead at the hospital. Employees #2 and #3 were admitted to the hospital for observation.

https://www.osha.gov/ords/imis/accidentsearch.accident_detail?id=200033223

August 2009 OSHA Report - Steel Dynamics (1 Fatality, 1 Serious Injury)

On August 8, 2009, Employee #1, who was a utility worker at a steel processing facility, went to the hot bridal portion of an annealing furnace to drop the ball and chain. Normally, when this task was assigned, the furnace was shut down for several days, and the nitrogen was locked out. However, on that day, the furnace was being rebricked and the nitrogen was left on to help the bricks cure. Fifteen minutes into the shift, at approximately 9:15 p.m., employees found Employee #1 unconscious, with the upper half of his body inside the hot bridle opening. Employee #2, who was the first to reach Employee #1, grabbed Employee #1 at the waist and pulled him out of the hot bridle opening. Employee #2 was then overcome by nitrogen. Other employees pulled Employees #1 and #2 away from the hot bridle opening. Employee #2 was sent to the hospital and made a full recovery. Employee #1 never responded to CPR or treatment at the hospital. He was pronounced brain dead and apparently died on August 8, 2009.

https://www.osha.gov/ords/imis/accidentsearch.accident_detail?id=200997880

October 2010 OSHA Report - Blowout Tools, Inc. (1 Fatality)

On October 30, 2010, Employee #1, a production supervisor, had installed a liquid nitrogen jacket around the outside of the conductor pipe and was flowing liquid nitrogen in order to form a plug. Employee #1 was found by coworkers, collapsed in the hole. He was removed by his coworkers, but he had already died.

https://www.osha.gov/ords/imis/accidentsearch.accident_detail?id=200212843

February 2011 OSHA Report - Enterprise Products Transportation Company, LLC (1 Fatality)

On February 7, 2011, Employee #1, a tank wash lead man, was cleaning the interior and exterior of a tank truck (Brenner Tank, Model SS Transport, VIN 10BFV7212NF0A3251) in wash bay 1 at Enterprise Products Transportation Company, LLC. He opened the manway and connected a compressed air hose to the rear washout coupling to ventilate the tank. At approximately 6:30 p.m., Employee #1 was found unresponsive inside the tanker truck immediately beneath the main manway hatch by an assistant tank washer working in wash bay 2. An alarm was sounded, and a coworker entered the tank and removed Employee #1. He was transported by ambulance to a nearby community hospital and later airlifted to a Houston hospital. Employee #1 died the next day. The product reported to have last been in the tanker truck was Dow Styrofoam(TM) RS 2030 Polyol. This chemical was a urethane-based product that was a mixture of 9 or more chemicals, including a number of oxirane, amine, oxide, formaldehyde and phthalate molecular compounds. MSDS information indicated Dow and ACGIH exposure limits as low as TWA of 0.05 ppm (skin) and STEL 0.15 ppm (skin). The MSDS information also indicated that in confined areas, the vapor can easily accumulate and cause unconsciousness and death due to displacement of oxygen (vapor pressure 27.2 psi & 32 deg C). The chemical was off-loaded from the tank trailer at a customer site using nitrogen. Residual nitrogen in the tank, combined with the chemical's oxygen displacing characteristics, could produce an oxygen-deficient atmosphere at the manway opening. The employees who rescued Employee #1 stated that a water hose used to test the product inside the tankers - to see if cold water washing would remove the product - was found running and dangling on the side of the tanker, and the Employee #1's flashlight was found inside the tanker. Apparently, Employee #1 positioned himself over the open manway to peer inside the tanker, was overcome, and fell into the tank.

https://www.osha.gov/ords/imis/accidentsearch.accident_detail?id=201925377

April 2011 OSHA Report - Ancien Wines Inc (1 Fatality)

At approximately 7:06 p.m. on April 20, 2011, Employee #1, an assistant wine maker at with Ancien Wines Winery, was transferring red wine from a small portable tank to a larger wine tank (10 ft high and 5 ft in diameter).The procedure included purging the larger tank and displacing oxygen with argon gas to prevent wine spoilage. Nitrogen gas was also used to inflate the rubber seal of the tank's floating lid after the wine transfer was completed. When Employee #1 failed to return his service truck to the employer's house after regular working hours, the employer tried unsuccessfully to reach Employee #1's cell phone. The employer then proceeded to the winery and found Employee #1 unconscious inside the tank. He died of asphyxia.

https://www.osha.gov/ords/imis/accidentsearch.accident_detail?id=201023314

February 2012 OSHA Report - Key Energy Services, Inc. (1 Fatality)

On February 16, 2012, Employee #1 was pumping nitrogen into an oil well. The company reported that Employee #1 was found unresponsive in the dog house while operating the equipment to pump nitrogen into the well. Employee #1 died from unknown causes.

https://www.osha.gov/ords/imis/accidentsearch.accident_detail?id=6455.015

June 2012 OSHA Report - ATI Allvac (1 Fatality)

On June 26, 2012, Employee #1 was working as a heat treat operator at a metal alloy manufacturing facility. He and a coworker, who was also working as a heat treat operator, were monitoring the progress of steel rectangles that were cooling in a cryogenic nitrogen box. The cryogenic nitrogen box was used to supercool metal at a controlled rate. The thermocouple in the back right corner was giving readings outside the acceptable range of 0 to 30 degrees. Employee #1 went out to the nitrogen box to service the thermocouple that was reading approximately minus 40 degrees. The narrative did not state whether these measurements were in degrees Fahrenheit or degrees Celsius. The nitrogen box was constructed of steel and plywood. It had been designed by ATI Allvac in 1998. The nitrogen box was approximately 26 feet in length; 6 feet, 6 inches in width; and 4 feet, 10 inches in height, with the lid closed (7.92 meters long, 1.98 meters wide, and 1.47 meters high). When the lid was open, workers had to step up 17 inches (0.43 meters) into the box. The size of the opening that employees used to enter the box was approximately 26 feet (7.92 meters) in length and 6 feet, 1 inch (1.85 meters) in height. The internal width of the box was approximately 5 feet, 6 inches (1.68 meters). There were a total of seven risers inside the box. Each riser was approximately 10 inches (254 millimeters or 25.4 centimeters) in height. The metal walking surface inside the nitrogen box was icy and slippery, and the area was poorly lit. The nitrogen box's 3,800-pound (1,724-kilogram) lid was lifted by two Hydro-Line single-acting standard pneumatic cylinders on the right and left sides of the box. The cylinders had a 1.375-inch rod diameter, a 40 inch stroke, and a 6 inch bore. (3.5-centimeter rod diameter, 101.6-centimeter stroke, and 15.2-centimeter bore). The pressure of the air going through the flexible air hose was 98 psi (676 kilopascals or 6.76 bar) to105 psi (724 kilopascals to 7.24 bar). The flexible air hose, rated at 200 psi (1,379 kilopascals or 13.79 bar) was attached to a hard line using a brass coupling and an HS10 Dixon worm gear clamp rated for used up to 30 psi (207 kilopascals or 2.07 bar). The brass coupling connecting the hard line to the flexible air hose was rated at 150 psi (1,034 kilopascals or 10.34 bar). According to the manufacturer, the worm gear clamp was not adequate to hold the hose in place at a pressure of 98 psi (676 kilopascals or 6.76 bar). Employee #1 went inside the nitrogen box to reposition and secure the thermocouple. Employee #1 did not engage the safety arms while adjusting the thermocouples on the steel rectangles inside the box. As Employee #1 was inside the box, the hose clamp failed, and the lid closed. Employee #1 was asphyxiated due to nitrogen inhalation, according to the coroner. Secondary injuries were related to extreme cold, but they were not the cause of death. Two maintenance workers found Employee #1 30 minutes to 1 hour after the incident had occurred. There were four violations of OSHA standards related to the accident. Citations were also issued for not developing machine-specific lockout procedures and for not engaging the safety arms. An additional citation was issued for failing to install locking pins or other mechanisms to secure the lid of the nitrogen box in place. A final citation was issued for not training employees on the procedures for isolating the nitrogen box.

https://www.osha.gov/ords/imis/accidentsearch.accident_detail?id=200376119

July 2012 OSHA Report - A C & S, Incorporated (1 Fatality)

On June 25, 2012, an employee was preparing to sand blast steel pylons on unit 2A and 2B. A coworker was instructed by an authorized sandblaster to hook up the air hose across the road. The coworker hooked up the airline to a nitrogen line instead of a supplied airline.

The employee donned the supplied air hood and lost consciousness due to asphyxiation from the nitrogen and died.

https://www.osha.gov/ords/imis/accidentsearch.accident_detail?id=20515.015

October 2013 OSHA Report - Vessel Repair, Inc (1 Fatality)

At 5:15 p.m. on October 27, 2013, three employees and a supervisor were conducting work on the deck of three barges to ready them for delivery. The supervisor and one of the employees made entry into the #1 cargo tank of barge's one and two to check the deep well blind on the deep well inspection flange. Another coworker assisted by Employee #1 also entered the #1 cargo tank. Employee #1 became dizzy and could not climb the ladder to get out, and fell to the tank bottom, landing on his right side and striking his head on the metal tank. Employee #1 was seen by his coworker bleeding from his head and unresponsive. The coworker notified the supervisor, who called the shipyard project coordinator who in turn called 911. The project coordinator told the supervisor that the barge entered by Employee #1 had nitrogen in all of its cargo tanks in preparation for receiving cargo. The fire department arrived on scene within 15 minutes of the initial fall from the ladder by Employee #1. The fire department tested the atmosphere of the tank and found there to be less than 1 percent oxygen. Employee #1 was asphyxiated and killed.

https://www.osha.gov/ords/imis/accidentsearch.accident_detail?id=53926.015

February 2014 OSHA Report - Qc Of Kentucky Inc (1 Fatality)

On February 7, 2014, Employee #1, a Tank Cleaner, was getting ready to work on a tractor trailer tanker. The employee dropped his flashlight into the tanker and went into the tanker to retrieve it. The tanker had just been emptied of Gantrez S-97, which is a copolymer used in toothpaste. The product had been purged from the tanker by pumping Nitrogen into the tanker which would help remove the product. The company believed the Nitrogen purge reduced the oxygen to dangerously low levels and Employee #1 died of chemical poisoning.

https://www.osha.gov/ords/imis/accidentsearch.accident_detail?id=202570636

December 2015 OSHA Report - SPX Transformer Solutions, Inc. (2 Fatalities, 1 Hospitalized)

On November, 30, 2015, employee #1 entered a large transformer, top cover removed, that had previously been filled with nitrogen. Employee #1 fell unconscious. Employee #2 entered to retrieve Employee #1 and also fell unconscious. Employee #3 followed and also fell unconscious. Employees #1 and #2 died. Employee #3 was hospitalized.

https://www.osha.gov/ords/imis/accidentsearch.accident_detail?id=83095.015

February 2016 OSHA Report - Tradebe Treatment and Recycling LLC (1 Fatality)

On February 19, 2015, employees were to perform preventive maintenance inspection inside of a scrubber, which is maintained free of oxygen by means of a nitrogen blanket. When the designated attendant could not find the entrant employee, other employees searched for him. The entrant employee was found inside the scrubber (a metal tube) where he had been preparing to clean it out. He was found to be unresponsive. The employee was pronounced deceased at the scene by the Lake County coroner.

https://www.osha.gov/ords/imis/accidentsearch.accident_detail?id=83115.015

May 2017 OSHA Report - Richey Collision Center, Inc (1 Fatality)

At 5:15 p.m. on May 16, 2017, an employee was preparing to paint a car. The employee hooked up his airline respirator to a nitrogen hose instead of to the compressed air hose, resulting in the employee suffering asphyxiation.

https://www.osha.gov/ords/imis/accidentsearch.accident_detail?id=95669.015

December 2017 OSHA Report - Union Tank Car Company (1 Fatality)

At 9:00 a.m. on December 15, 2017, an employee was cleaning the excess oil from the bottom of a tanker rail car following a previous application of oil to prevent corrosion. The employee collapsed due to a 98% atmosphere of nitrogen being present inside the tank car. The employee was killed due to asphyxiation.

https://www.osha.gov/ords/imis/accidentsearch.accident_detail?id=101665.015

July 2018 OSHA Report - Barry Callebaut USA LLC (Hospitalized Injury)

At 2:00 p.m. on July 17, 2017, Employee #1, a maintenance mechanic, was working in Building A, at the Black Powder Line 3. The employee opened the upper access hatch on the vessel that contained the Cake Mill Air Heat Exchanger, to replace an old air filter that was located on top of the heat exchanger. During normal operations, nitrogen gas from a near-by cylindrical shape metal tank, flows through the heat exchanger. The employer used nitrogen as an inert gas to remove oxygen from the process equipment to decrease the chance of a fire or explosion. Employee #1 was working alone and later, a coworker discovered him face down with his face, chest, shoulders, arms, and hands inside the heat exchanger. The coworker saw him through the opening of the upper hatch of the vessel and the employee was unconscious and unresponsive. His coworkers pulled him out of the heat exchanger, and the paramedics transported him to a nearby medical center. He remained in the hospital for treatment of nitrogen.

https://www.osha.gov/ords/imis/accidentsearch.accident_detail?id=109440.015

November 2018 OSHA Report - Krones Inc / Krones Ag (2 Serious Injuries)

At 8:00 a.m. on November 8, 2018, Employee #1 was performing mechanical support duties. The accident involved a Decontamination Module, Identified by the Employer as KA30011. Employee #1 was assigned to change metal plates (15 inch by 2 inch, approximately) located inside the Denomination Unit, a permit-required confined space. Employee #1 gathered the tools needed to replace the metal plates and climbed up to the location of the confined space entry. The employee went through the

18 inch (approximately) opening and began descending the 16 rungs (approximately) ladder, into a unit containing nitrogen gas. Consequently, within 3–5 seconds (approximately), Employee #1 lost consciousness due to the oxygen-deficient atmosphere and fell over 10 feet below. Soon after, Employee# 2, went through the opening and began to descend the ladder when he suffered the same effects and fell onto Employee# 1. Employee# 1 and Employee# 2 were transported to LAC + USC Medical Center, where they were treated for their (note: OSHA log incomplete).

https://www.osha.gov/ords/imis/accidentsearch.accident_detail?id=115734.015

June 2019 OSHA Report - Terry And Sons, Inc. (1 Fatality)

At 8:30 a.m. on June 15, 2019, Employee #1, employed by a painting contractor, was setting up equipment to abrasive blast electrode holders. He connected the abrasive blasting hood to a line that was supplying nitrogen rather than one supplying compressed air. Employee #1 inhaled nitrogen instead of breathable air and was asphyxiated.

https://www.osha.gov/ords/imis/accidentsearch.accident_detail?id=117257.015

November 2019 OSHA Report - Custom Genetic Solutions, LLC (1 Fatality)

At 8:00 p.m. on November 20, 2019, an employee was topping off bull semen storage tanks with liquid nitrogen in a tank storage room when the nitrogen evaporated and displaced the oxygen in the room. The employee was killed from asphyxiation while working in a non-ventilated room.

https://www.osha.gov/ords/imis/accidentsearch.accident_detail?id=122100.015

February 2020 OSHA Report - Kenan Advantage Group, Inc. / Great Lakes Tank & Vessel LLC (2 Fatalities)

At 12:30 p.m. on February 20, 2020, Employee #1, employed by a structural steel fabricator and erector company, was entering a tank to clean it. The tank had a combination of Ecocure II and methyl ethyl ketone (MEK) residues and had been purged with nitrogen. Employee #1 entered the permit required confined space that contained the residual chemicals and nitrogen to perform the cleaning operations. She was overcome by the oxygen deficient atmosphere. Employee #2, employed by a chemical distribution company, entered the tank to make a rescue attempt for Employee #1. He was also overcome from the oxygen deficient atmosphere. Both employees were killed by asphyxiation.

https://www.osha.gov/ords/imis/accidentsearch.accident_detail?id=124207.015

May 2020 OSHA Report - Steel Line Rail Services LLC (1 Fatality)

At 9:30 a.m. on May 12, 2020, Employee # 1 and a coworker were monitoring the tank atmosphere with a gas meter. The coworker entered and exited the tank immediately, while Employee # 1 went inside the tank. Employee # 1 was overcome with nitrogen gas or oxygen deficiency. He was killed.

https://www.osha.gov/ords/imis/accidentsearch.accident_detail?id=126735.015

July 2020: Research Gate reported four cases during July 2020 (4 Fatalities)

We report four cases of fatal poisoning due to nitrogen gas. Three of them died after penetrating without personal protective equipment, in a tank containing nitrogen gas, deaths occur within a few hours to five days after the accident. The fourth case occurred when a worker penetrated a tank that had contained tetrachloroethylene and had been replaced by nitrogen gas. The worker was rescued 15 minutes later by co-workers, is in a deep coma and goes into a vegetative state and died six weeks after the accident

https://www.researchgate.net/publication/346963946_Occupational_Deaths_Due_to_Nitrogen_Gas

December 2020 OSHA Report - California Ranch Food Company, Inc. (2 Fatalities)

At 7:45 p.m. on December 1, 2020, Employee #1, a production lead supervisor, entered the chill/production room. She observed a cloud generating in the immediate area of the freezer tunnel. She went to investigate the cause of the cloud and was killed by asphyxiation. Approximately one hour later, Employee #2, a production supervisor, entered the chill/production room. He observed Employee #1 on the floor and went over to her. Employee #2 was overcome by the lack of oxygen in the room and was also killed by asphyxiation. Liquid nitrogen released from the tunnel freezer, Linde Cryowave Freezing tunnel Model 1250-5, in the chill/production room created an oxygen-deficient atmosphere that asphyxiated both employees. The chill/production room emergency alarms, Advanced Micro Instruments 221R Area Oxygen Deficiency monitor, did not provide warning to the employees of an oxygen-deficient atmosphere.

https://www.osha.gov/ords/imis/accidentsearch.accident_detail?id=131418.015

January 2021 OSHA Report - Foundation Food Group, Inc. (6 Fatalities)
(Also see CSB Report on this incident)

At 10:13 a.m. on January 28, 2021, three maintenance employees were in a freezer room troubleshooting improperly frozen cooked chicken parts discharging from the liquid nitrogen Messer Immersion and Spiral Freezer. The immersion freezer and the internal freezer tub were designed to contain a bath of liquid nitrogen to flash freeze the chicken. One of the maintenance employees, who was troubleshooting the freezer, used a tool to bend the bottom of the stainless steel bubbler system dip tube upward inside the tub of the immersion freezer. The bubbler system was part of the liquid control system that maintained the level of liquid nitrogen in the immersion freezer tub. Soon after the freezer was restarted, liquid nitrogen in the immersion freezer tub overflowed onto the floor due to the damaged dip tube. The liquid nitrogen that spilled out of the freezer converted to vapor due to the changes in temperature and pressure and displaced oxygen from the air. The three maintenance employees succumbed to the oxygen deficient atmosphere inside the freezer room. The plant superintendent, a maintenance employee, and a quality control technician went to the freezer room and attempted to help the employees when they also died due to the oxygen (OSHA report missing words).

https://www.osha.gov/ords/imis/accidentsearch.accident_detail?id=132799.015

February 2021 OSHA Report - Freedomtrucks of America, LLC (1 Fatality)

At 9:45 a.m. on February 22, 2021, an employee was working for a general, long-distance, truckload freight trucking company. He washed vehicles and cleaned equipment for the company. He was cleaning a tanker trailer that had contained ALFOL 16. This is a tradename for 1-Hexadecanol. The employer stated that the employee was inspecting the tank for materials. The atmosphere in the tanker had been charged with nitrogen. It was low in oxygen. The employee lost consciousness and died from asphyxia. He was not hospitalized.

https://www.osha.gov/ords/imis/accidentsearch.accident_detail?id=133632.015

August 2021 OSHA Report - Knapheide Truck Equipment Company Midsouth (1 Fatality)

At 12:00 noon on August 19, 2021, an employee utilized an airline respirator which was attached to a nitrogen line. The employee died from lack of oxygen.

https://www.osha.gov/ords/imis/accidentsearch.accident_detail?id=138501.015

May 2022 OSHA Report - Northrop Grumman Corporation (1 Fatality)

At 8:38 a.m. on May 19, 2022, an employee purged the Nitrogen lines of the battery freezers in a lab room and the freezers filled up with Nitrogen gas. The victim closed himself into one of the freezers, where he eventually died from the overexposure to Nitrogen gas and lack of oxygen.

https://www.osha.gov/ords/imis/accidentsearch.accident_detail?id=148226.015

February 2023 OSHA Report - Specialty Welding And Turnarounds, LLC (1 Fatality)

At 6:45 a.m. on February 18, 2023, an employee was removing trays from a hydrogen reactor. The employee was on supplied breathing air. The employee was attempting to exit the confined space at the end of the shift, when the employee reported to the hole watch that he could not breath. The hole watch told the employee to turn on the emergency bottle. The employee then reported that he was on the bottle. The hole watch told the employee to connect to the emergency escape line. The employee refused and attempted to climb out of the confined space. The employee climbed approximately 30 feet up a fixed vertical ladder. The employee stopped responding to the attendant and other exterior employees. A man down alarm was called, and emergency rescue operations were attempted. The employee was extracted from the confined space. CPR was performed. An AED was applied, and the AED advised not to administer a shock. The employee was transported by ambulance to a nearby hospital emergency department, where he was pronounced dead. The employee was killed by probable asphyxia due to nitrogen gas exposure and/or oxygen deprivation.

https://www.osha.gov/ords/imis/accidentsearch.accident_detail?id=153944.015

May 2023 OSHA Report - Energy Contract Services, LLC (1 Fatality)

At 8:04 a.m. on May 11, 2023, an employee working for an oil and gas support services company was found dead. On 5/10/23, the night before, the employee had instructed his coworkers from other contractors to go home, and he would finish the job up. This was the last time anyone had spoken to the employee until he was found deceased. The employee had been pressure testing a pipe with nitrogen. The pipe had been opened and nitrogen was leaking into the ambient environment. The employee was asphyxiated due to the displacement of oxygen causing his (OSHA log incomplete).

https://www.osha.gov/ords/imis/accidentsearch.accident_detail?id=160306.015

Index

Note: Page numbers in *italics* and **bold** refers to figures and tables, respectively.

a

air-purifying respirator, 108
Alaska Welders Helper argon asphyxiation, 152–155
argon (Ar)
    Alaska Welders Helper asphyxiation, 152–155
    description of, 14
    Freeport-McMoRan Morenci Inc. incident, 163–164
    health hazards of, 14
    oxygen deprivation, 25–27
    precautions for storage and handling cylinders, 14–15
    shipping information, 15
    Singapore worksite incident, 155–158, *156*
asphyxiation
    actions to prevent incidents, 171–172
    Alaska Welders Helper argon asphyxiation, 152–155
    argon, 152–158, *156*, 163–164
    carbon dioxide, 159–160
    C&M Roustabout Services LLC incident, 160–161
    Foundation Food Group incident, 144–148, *145*, *146*
    Freeport-McMoRan Morenci Inc. incident, 163–164
    helium, 18–19
    Kentucky ice cream facility incident, 163
    krypton, 21
    light hydrocarbons, 22
    Meriwether Compressor Station incident, 161–163
    nitrogen. *see* nitrogen asphyxiation
    rescuers, death of, 164–168
    Singapore worksite incident, 155–158, *156*
    Smoley Mountain Opry Music Venue incident, 159–160
    Valero Delaware City Refinery incident, 149–152, *149–151*
    xenon, 21

b

breathing air (BA)
    blended or manufactured, fatalities due to, 96
    cylinder, 107
    instrument air as, 95–96
    quality, assuring, 125–126
    requirements for, 96–97
    system, 122–125, *122–125*
    utility air as, 95–96

c

carbon
    capture, 134–136
    storage, 137
    transport, 136–137
Carbon Capture, Utilization and Storage (CCUS), 133

Hazards of Nitrogen and Other Inert Gases: How They Can be Safely Managed, First Edition. M. Darryl Yoes.
© 2025 John Wiley & Sons, Inc. Published 2025 by John Wiley & Sons, Inc.

Carbon Capture, Utilization and Storage (CCUS), (*continued*)
  design and construction, 137–138
  industry response to increase in carbon dioxide, 134–137
  safety aspects of, 137
  start-up and operation, 138
  storage or sequestration, 138
  transport, 138
carbon dioxide (CO_2)
  description of, 15
  and greenhouse gas elimination, 15–16
  health hazards of, 15, **16**
  industry response to increase in, 134–137
  oxygen deprivation, 25–27
  Smoley Mountain Opry Music Venue incident, 159–160
carbon monoxide (CO), 13, 16–17
  oxygen deprivation, 25–27
cardinal safety rules, 104
catalyst crusting, 129–130, *129*, *130*
Chemical Safety and Hazard Investigation Board (CSB), 3, 32–33, 143, 145
chemical solvents, 135
C&M Roustabout Services LLC incident, 160–161
cold nitrogen, 25
confined space entry, 38, *38*, *39*
  Freeport–McMoRan Morenci Inc. incident, 163–164
  and inert entry rescue plan, 127
  nitrogen asphyxiation in, 108–109
  Workplace Safety & Health Council (WSH) Regulations, 157
contractor selection process, 126–127
cryogenic fractionation, 136
cryogenic gases, transport and storage of, 6–7

d
de minimis violation, 80

e
emergency medical technicians (EMTs), 154–155
Environmental Protection Agency (EPA), 1–2

f
food processing industry, liquid nitrogen in, 7–8

Foundation Food Group incident, 144–148, *145*, *146*
Freeport–McMoRan Morenci Inc. incident, 163–164

g
greenhouse gas, elimination of, 15–16

h
health hazards
  of argon, 14
  of carbon dioxide, 15, **16**
  of helium, 19
  of krypton, 20
  of liquid nitrogen, 3
  of neon, 20
  of nitrogen, 2–3
  of xenon, 21
helium (He)
  asphyxiation, 18–19
  characteristics and hazards, 18
  description of, 19
  health hazards of, 19
  overview, 17–18
  oxygen deprivation, 25–27
  precautions for safe handling and storage, 19
  shipping information, 19
  uses of, 18

i
immediately dangerous to life and health (IDLH) environment
  breathing air (BA) cylinder, using, 107
  running out of air in SCBA, 108
  with wrong type of respirator, 108
inert entry
  acceptable atmosphere, 121
  background, 118–119
  breathing air quality, assuring, 125–126
  breathing air system, 122–125, *122*, *123*, *124*, *125*
  catalyst crusting, 129–130, *129*, *130*
  caution, 128–129
  confined space and rescue plan, 127
  contractor selection process, 126–127
  planning, 120

hazards of, 117–130
inert gas supply and quality, 122
job safety analysis, 121
life support equipment, 127
reactor preparation for, 128
specialized procedures, 119–120
video surveillance/rescue equipment, 127
inert gases
argon, 13–15
carbon dioxide, 15–16, **16**
carbon monoxide, 16–17
helium, 17–19
krypton, 20
light hydrocarbons, 21–22
neon, 19–20
personnel protection against asphyxiation, 31–40
supply and quality, 122
xenon, 21
instrument air, use of, 95–96

j
job safety analysis (JSA), 121

k
Kentucky ice cream facility incident, 163
krypton (Kr)
asphyxiation, 21
characteristics and uses of, 20
description of, 20
health hazards of, 20
oxygen deprivation, 25–27

l
light hydrocarbons, 21–22
oxygen deprivation, 25–27
suffocation (asphyxiant) hazard, 22
liquid nitrogen
cylinders, transporting, 109
explosion while unloading at ice cream facility, 163
in ice cream shops, 175–177
individual worker, 40
management, 39–40
protection against, 39, *39*
training, 40

m
membrane systems, 135
Meriwether Compressor Station incident, 161–163

n
neon (Ne)
description of, 20
health hazards of, 20
oxygen deprivation, 25–27
uses of, 19–20
nitrogen. *see also* liquid nitrogen
as gas, 7, 8
in food processing industry, 7–8
health hazards of, 2–3
incidents resulting in fatalities or serious injury, 219–250
oxygen deprivation, 25–27
in petrochemicals industry, 7
in petroleum refining, 7
properties of, 1–8
safety hazards of, 2
safe utility connections, 111–114, *112*, *113*
uses of, 7–8
nitrogen asphyxiation
in confined space, 108–109
Foundation Food Group incident, 144–148, *145*, *146*
IDLH environment, working in, 107–108
nitrogen cylinders, transporting, 109
respirators and/or sandblast hood, connecting, 103–104, *105*
storage tank, entering, 104–106
Valero Delaware City Refinery incident, 149–152, *149–151*
without proper emergency response and PPE, 106
working near open process vessel, 106–107

o
Occupational Safety and Health Administration (OSHA), 3–4, 17, 45, 104, 143
breathing air quality, requirements for, 96–97, **97**
confined space entry letters of interpretation by, 64–92
confined space regulation, 45–63
Technical Manual on Respiratory Protection, 96–97

oxy combustion, 136
oxygen concentration, 121
oxygen deprivation, 25–27. *see also* asphyxiation
   C&M Roustabout Services LLC incident, 160–161
   Meriwether Compressor Station, 161–163
   physiological effects of, **26**

p

permit-required confined space (PRCS) definition, 65–66
Personal Alert Safety System (PASS) alarm, 108
personal protective equipment (PPE), 118–119
personnel protection
   adequate warning signs and barricades, 31–33, *32*, *33*
   confined space entry, 38, *38*, *39*
   Delaware City Refinery Incident, 36–38, *37*
   documented work permit, 36
   gas monitors and continuous gas quality monitoring and alarms, 33–35, *34*
   protection against supercold liquids, 39–40, *39*
   self-contained breathing apparatus, use of, 35–36, *35*
petrochemicals industry, nitrogen in, 7
petroleum refining, nitrogen in, 7
process hazard analysis (PHA), 138

r

rescuers, death of, 164–168

s

safe utility connections, 111–114, *112*, *113*
safe work procedures (SWPs), 156
self-contained breathing apparatus (SCBA)
   running out of air in IDLH environment, 108
   use of, 35–36, *35*
Singapore worksite incident, 155–158, *156*
single time exposure limits (STEL), 76–77
Smoley Mountain Opry Music Venue incident, 159–160
sorbents, 135
suffocation. *see* asphyxiation

t

Tennessee Occupational Safety and Health Administration (TOSHA), 159–160

u

utility air
   as breathing air, 95–96
   utility connections safety, 111–114, *112*, *113*

v

Valero Delaware City Refinery incident, 149–152, *149–151*

w

whip check, 111, *113*

x

xenon (Xe)
   asphyxiation, 21
   description of, 21
   health hazards of, 21
   oxygen deprivation, 25–27